Ad Hoc Networks Telecommunications and Game Theory

FOCUS SERIES

Series Editor Guy Pujolle

Ad Hoc Networks Telecommunications and Game Theory

Malek Benslama
Mohamed Lamine Boucenna
Hadj Batatia

iSTE

WILEY

First published 2015 in Great Britain and the United States by ISTE Ltd and John Wiley & Sons, Inc.

ISTE Ltd
27-37 St George's Road
London SW19 4EU
UK

www.iste.co.uk

John Wiley & Sons, Inc.
111 River Street
Hoboken, NJ 07030
USA

www.wiley.com

Library of Congress Control Number: 2014955867

British Library Cataloguing-in-Publication Data
A CIP record for this book is available from the British Library
ISSN 2051-2481 (Print)
ISSN 2051-249X (Online)
ISBN 978-1-84821-774-4

Contents

FOREWORD ix

INTRODUCTION xi

LIST OF ACRONYMS xv

CHAPTER 1. *AD HOC* NETWORKS: STUDY AND DISCUSSION OF PERFORMANCE 1

1.1. Introduction 1
1.2. Concepts specific to *ad hoc* networks 2
1.2.1. Topology 2
1.2.2. Connectivity 3
1.2.3. Mobility 4
1.2.4. Networks: wireless mesh network (WMN), wireless sensor networks (WSN) and mobile *ad hoc* network (MANET) 5
1.2.5. Routing 7
1.2.6. Weak security 9
1.2.7. Access to the environment 9
1.3. MAC protocols in mobile *ad hoc* networks 10
1.3.1. ALOHA 10
1.3.2. CSMA 15
1.4. Energy consumption in *ad hoc* networks 25
1.4.1. Energy overconsumption and/or waste 28
1.4.2. Toward more efficient energy consumption 30
1.5. Conclusion 34

Chapter 2. Game Theory and Communication Networks 35

2.1. Introduction 35
2.2. Introductory concepts in game theory 38
2.2.1. Game 38
2.2.2. Player 38
2.2.3. Strategy (pure and mixed) 39
2.2.4. Utility 40
2.2.5. General classification of games 41
2.2.6. Equilibrium 44
2.3. Nash equilibrium 46
2.3.1. Definition 46
2.3.2. Existence 47
2.3.3. Uniqueness 50
2.3.4. Specific cases 51
2.4. Famous games 52
2.4.1. The prisoner's dilemma 52
2.4.2. Cournot duopoly 53
2.5. Applications to wireless networks 55
2.5.1. Routing game 56
2.5.2. Power control game 58
2.6. Conclusion 60

Chapter 3. Games in SALOHA Networks 61

3.1. Introduction 61
3.2. Functioning of the SALOHA algorithm 64
3.2.1. Study of stability 68
3.2.2. Transmission time 72
3.3. Modeling of node behavior in SALOHA with a strategic coding game 73
3.3.1. Issues 73
3.3.2. RS erasure codes 75
3.3.3. The impact of erasure encoding on SALOHA 79
3.3.4. Description of game model 80
3.3.5. Study of utility 83
3.3.6. Discussion of equilibrium 84
3.4. SALOHA network performance at Nash equilibrium 86
3.4.1. Coding cost 86

3.4.2. Loss rate 87
3.4.3. Output 88
3.4.4. Stability 90
3.4.5. Transmission time 91
3.5. Conclusion 92

CHAPTER 4. GAMES IN CSMA NETWORKS 93

4.1. Introduction 93
4.2. CMSA performance 95
4.3. Sources of problems in CSMA networks 99
4.4. Modeling of node behavior in CSMA using a strategic coding game 100
4.4.1. Game model analysis 100
4.4.2. Utility function 101
4.4.3. Discussion of equilibrium 103
4.5. CSMA performances at equilibrium 105
4.5.1. Coding/decoding price (cost) 105
4.5.2. Output 106
4.5.3. Transmission time 108
4.5.4. Energy optimization at equilibrium 109
4.6. Conclusion 110

CONCLUSION 113

BIBLIOGRAPHY 119

INDEX 139

Foreword

Four books devoted solely to satellite communication: this was the challenge laid down by Professor Malek Benslama of the University of Constantine, who understood that a new discipline was in the process of taking shape.

He demonstrated this by organizing the first International Symposium on Electromagnetism, Satellites and Cryptography in Jijel, Algeria, in June 2005. The success of this conference, which was surprising for an inaugural event, demonstrated the need for specialists with skills that sometimes varied widely from one another to come together in the same place. The 140 papers accepted concerned not only systems but also electromagnetism, antenna and circuit engineering, and cryptography, which often falls under the category of pure mathematics. Synergy must exist among these disciplines in order to develop the new field of activity that is satellite communication.

We have seen new disciplines of this type emerge in the past; for electromagnetic compatibility, it was necessary to understand both electrical engineering (for guided modes and choppers) and electromagnetism (for propagated modes) and to know how to define specific experimental protocols as well. Further back in time, computer science was the domain of electronics engineers in its early days, and became a separate discipline only gradually.

Professor Benslama has the knowledge and open-mindedness needed to combine all the areas of expertise that coexist in satellite telecommunications. I have known him for 28 years now, and it has been a real pleasure for me to look back on all those years of acquaintance. Not a single year has gone by when we have not seen each other. He spent the first 15 years of his career working on the interaction between acoustic waves and semiconductors, specializing in the solution of piezoelectric equations (Rayleigh waves, surface skimming waves, etc.) while taking an interest in theoretical physics at the same time. A PhD degree in engineering and, later, a high-level State doctorate degree were added to his many achievements. Among the members of his dissertation committee was Madame Hennaf, then Chief Engineer at CNET (the National Centre for Telecommunications Studies in Issy Les Moulineaux). He had already developed an interest not only in telecommunications but also, with the presence of Monsieur Michel Planat, head of research at the National Centre for Scientific Research at LPMO Besançon (CNRS), in the difficult problem of the synchronization of oscillators.

With Michel Planat, he embarked on the path that would lead him to quantum cryptography, a conversion that he has made over the past 10 years, passing without apparent difficulty from Maxwell equations to Galois groups.

He is now one of the people most capable of mastering all the diverse disciplines that form satellite telecommunications.

I hope, with friendly admiration, that these four monographs will receive a warm welcome from both students and instructors.

Professor Henri BAUDRAND
Professor Emeritus
ENSEEIHT Toulouse
November 2014

Introduction

The first approaches concerning game theory date from the years 1921–1927 and were made by Emile Borel [BOR 21]. In 1928, J. Von Neumann introduced what he called "Zur Theorie der Gesellschaftsspiele" [VON 28]. But it was only in 1944 that J. von Neumann and O. Morgenstern applied game theory to the study of economic behavior [VON 44]. John Nash has studied non-cooperative games with applications in the field of economics [NAS 51, NAS 53].

Game theory has been dealt with extensively in the literature focused principally on economics and strategy [LUC 59, DRE 61, FUD 91, MYE 91, GIB 92, CAM 05, OSB 94, FUD 98, OSB 00, ALP 05, AGH 06, WAT 13, BRO 08, HAR 10, DIX 10, JUL 12, FOR 99].

The application of game theory to wireless and *ad hoc* networks dates from 1995, with Weibull the main precursor [WEI 95]. Algorithms specific to game theory have been posited [NIS 07]. Since 2008, the application of game theory to communication networks has been developed in [NIS 07, HAN 08, PAL 10, ZHA 11, HAN 12, DOR 14].

The production of the first radio transmission in 1896 by Guglielmo Marconi ushered in a new world of wireless telecommunication. This led to continual improvements in this world, due to new ideas and techniques proposed by scientists to facilitate and accelerate the act of communication. These days, because of the various services they offer, telecommunication resources have become essential in most of our daily activities – telephones, television, radio and the Internet; remote surveillance, control and detection; etc. These are services that we use frequently every day in various areas. Now, in addition to audiovisual communication services that connect people and shrink distances, the Internet allows users around the world to exchange data at speeds that are constantly evolving, and new means of telecommunication have also made it possible to ensure the safety of the environment via the use of remote detection and video surveillance techniques, as well as the setting up of small wireless networks that help people to escape quickly from possible injury or natural catastrophe. Moreover, developments in telecommunication have been highly conducive to the evolution of other important fields, notably delicate areas such as medicine and scientific research, as the latter, for example, uses sophisticated communication techniques to discover and analyze new biological and spatial phenomena.

Wireless networks are modern means of communication. Currently, they are very widely used in various aspects of life. A wireless network is composed of several stations, sometimes called nodes, which communicate with one another via electromagnetic wave-based radio links. New users can easily access and communicate via these networks without the need to install new infrastructures or cables. In addition to this advantage, wireless networks are less costly, easily deployable and possessed of dynamic topologies that enable node mobility, but in order to take full advantage of them, certain pitfalls caused by this mobility, such as service quality and security, must be overcome. However, users of a wireless network share a single communication channel to transmit their data; therefore, medium access control (MAC) is vital in order to avoid interference between the signals transmitted, thus ensuring stable and efficient functioning for an adequate period of time. To access the medium, nodes in wireless networks use the IEEE 802.11 distributed coordination function (DCF) protocol, which is an improved version of the basic carrier sense multiple access (CSMA) protocol, which is part of the family of random access protocols. The CSMA was preceded by the slotted ALOHA (SALOHA); both techniques are used to grant multiple accesses to the transmission support, but CSMA also possesses a channel-listening mechanism that can detect whether the channel is free to start the transmission; if it is not, it is necessary to wait for the channel to become free. It should be noted, however, that despite all the prevention techniques added to the CSMA protocol, collision still occurs and has a negative impact on the various performances of a wireless network.

The phenomena of interference and collision are indicative of the interactions and conflicts that exist between a network's users. All of this explains the importance of the application of game theory, the main objective of which is to move toward the efficient allocation of resources, energy control and optimization of output. The proof of this is that in recent years, the allocation of resources based on game theory has considerably improved effectiveness in the utilization of the radio spectrum. Even in terms of the physical (PHY) and MAC layers, the atmosphere seems quite well suited to the application of game theory. This is due mainly to the various conflicting situations encountered by these two layers, to the extent that in the MAC layer, users share the same channel to access the medium; the same routes to deliver packets; the same routers and sometimes the same transmitter and receiver nodes. Interactions in situations like this are confirmed and may negatively affect the network's yield. Nevertheless, the application of game theory to these types of circumstances is very useful in planning appropriate solutions that will lead to network stability and optimization.

The focus of this book falls within the same context, as we are presenting two models based on game theory to analyze the SALOHA and CSMA protocols, respectively. The model we are proposing in this book consists of a new idea that has not existed in the literature before now, and which involves the random use of the redundancy of an erasure coder to reduce collision and improve network performance in terms of output and transmission time. To this end, we have structured the book in the following way: we will begin Chapter 1 by presenting introductory concepts of wireless networks and their different characteristics, as well as a detailed study on random access protocols, and more precisely the CSMA protocol and its improved versions. We will end Chapter 1 by introducing the issue of energetic overconsumption, and we will explain the causes, circumstances and solutions proposed in the literature for the optimization of energy management in wireless networks, and more specifically mobile *ad hoc* networks (MANETs), which require additional quantities of energy to cover node mobility. Chapter 2 is devoted entirely to the presentation of game theory, its aims and the principal rules that govern it. We will conclude this chapter by discussing the existing approach between game theory and telecommunications, with explanatory examples.

Chapters 3 and 4 develop two game-coding models for SALOHA and CSMA, respectively. The goal of these two chapters is to demonstrate the possibility of optimizing output optimization, energy consumption and transmission time at the point of network convergence and equilibrium, given that at equilibrium, a game becomes stable and all nodes will be satisfied in terms of gain. This will have the effect of eliminating the desire of any node to change its strategy laterally. Finally, in order to assess our results and show the benefits they offer to random access networks, we have added several simulations to both chapters describing various network performances in terms of coding value, packet loss rate, output and transmission time, as well as the control of energy consumption.

List of Acronyms

ABR	Associativity-based routing
ACK	Acknowledgment
Ad hoc	Wireless network that does not rely on a pre-existing infrastructure
ALOHA	Random Access Protocol
AODV	*Ad hoc* on-demand distance vector
ARQ	Automatic repeat request
ATM	Asynchronous transfer mode
BER	Bit error ratio
CCA	Clear channel detection
CDMA	Code division multiple access
CEDAR	Core extraction distributed *ad hoc* routing
CRC	Cyclic redundancy check
CSMA	Carrier sense multiple access
CSMA/CA	CSMA with collision avoidance
CSMA/CD	CSMA with collision detection
CTS	Clear to send
DCF	Distributed coordination function
DIFS	Distributed inter frame space

DSDV	Destination sequenced distance vector
DSR	Dynamic source routing
DSSS	Direct sequence spread spectrum
ECIP	Erasure coding
EDCF	Enhanced DCF
EIFS	Extended IFS
FDMA	Frequency division multiple access
FEC	Forward error correction
FHSS	Frequency hopping spread spectrum
FIFO	First input first output
GAF	Geographic adaptive fidelity
GSM	Global system for mobile communication
GSR	Global state routing
HF	High frequency
IEEE 802.11	*Comité Internationale de Standardisation* (International Standardization Committee)
IFS	Inter frame space
IR	Infrared
LEO	Low earth orbit
LLC	Logical link control
MAC	Medium access control
MANET	Mobile *ad hoc* network
NE	Nash equilibrium
OFDM	Orthogonal frequency division multiplexing
OLSR	Optimized link state routing
OSI	Open systems interconnection
PAMAS	Power Aware Multi Access Protocol with signaling
PCF	Point Coordination Function

PCMCIA	Personal Computer Memory Card International Association
PLCP-PDU	Physical layer convergence protocol-Protocol data unit
PHY	Physical layer
PSM	Power-saving mode
QoS	Quality of service
RNG	Relative neighborhood graph
RPGM	Referenced point group mobility
RS	Reed–Solomon
RSSI	Received signal strength indicator
RTS	Request to send
SALOHA	Slotted ALOHA
SIFS	Short IFS
SNIR	Signal to interference plus noise ratio
SPAN	Switched port analyzer
S-MAC	Sensor-MA
SSA	Signal-stability adaptive routing
STEM	Sparse topology and energy management
TDMA	Time division multiple access
TORA	Temporally ordered Routing Algorithm
TS	Time slot
UDG	Unit disk graph
WiFi	Wireless fidelity
WLAN	Wireless local area network
WMN	Wireless mesh network
WSN	Wireless sensor network
WRP	Wireless Routing Protocol
ZHLS	Zone-based hierarchical link state
ZRP	Zone Routing Protocol

1

Ad Hoc Networks: Study and Discussion of Performance

1.1. Introduction

Ad hoc networks are wireless networks that form spontaneously and organize themselves automatically without requiring a pre-existing infrastructure. An *ad hoc* network is a collection of hosts (nodes) equipped with antennas that can communicate with one another without administrative centralization based on the topology of wireless communication. Unlike wired networks in which only a few nodes called routers are responsible for delivering data, in an *ad hoc* network, each node acts as a terminal mode, and possibly as a link to relay messages when recipients are not within radio range of the transmitters. In an *ad hoc* network, a node can communicate directly with another node in point-to-point mode when the two nodes are located in the same transmission zone, while communication with a node in another zone is carried out via several intermediary nodes in multi-hop mode.

Initially of military origin [JUB 87], due to several of the benefits they offer, *ad hoc* networks are of confirmed interest for circumstances characterized by a total lack of pre-existing infrastructures. From an applicative point of view, *ad hoc* wireless networks are useful in situations that require the deployment of a rapid local network or one lacking infrastructure, such as reaction to a crisis, conferences, military applications and possibly household and office networks. *Ad hoc* networks could, for example, enable medical personnel and officials to better coordinate their efforts during large-scale emergency situations that result in the complete destruction of network infrastructures, as was the case with the September 11th attacks and the 2003 blackout in the northeastern United States [JUR 07]. The principal advantages of *ad hoc* networks can be summed up in the following characteristics:

– Deployment: the deployment of an *ad hoc* network is quick, simple and low-cost.

– Organization: *ad hoc* networks are organized automatically via local collaboration between nodes with no centralization required.

– Adaptability: in the event of the breakdown of some nodes or links, *ad hoc* networks ensure continuity of operation and rebuild themselves locally using efficient access and routing techniques.

– Robustness: *ad hoc* networks ensure continuous operation and override critical states by executing local repairs without outside intervention.

Of the applications in which these types of networks are involved, we would note environmental applications such as the prevention of natural disasters and forest surveillance, as well as medical applications including the detection of physiological signals sensed on different parts of patients' bodies, etc. In these applications, the nodes are generally mobile and require specific power sources to be able to cover their energy needs while they are active. The issue of energy consumption is central here, and in the last part of this chapter, we will discuss it in detail, explaining the various solutions proposed in the literature, whose objective is to optimize and/or minimize energy consumption.

1.2. Concepts specific to *ad hoc* networks

1.2.1. *Topology*

An *ad hoc* network can be represented mathematically by an undirected graph $G = (V\ E)$, where V designates the group of nodes and $E \subseteq V^2$ denotes the group of arcs corresponding to the direct communications possible. If *u* and *v* are two nodes of *V*, l'arc $(,)u\ v$ exists if and only if *u* can send a message directly to *v*; in this case, we say that *v* is neighbors with *u*. Pairs belonging to *E* are dependent on the position of the nodes and their communication ranges. If we assume that all of the nodes have an identical range *R* and if $d\ u\ v(,)$ designates the distance between nodes *u* and *v*, then [CHE 98]:

$$E = \left\{(u,v) \in V^2 \middle| d(u,v) \leq R\right\} \qquad [1.1]$$

The graph thus obtained is called a unitary graph. If a node reduces its communication range, its relationships with the other nodes will be changed. This operation leads to a modification of the network's topology, which is expressed by a change of the unitary graph by another derived graph. The derived graph is extracted on the basis of known models, for example the unit disk graph (UDG) model with the relative neighborhood graphing (RNG) technique, the objective of which is to remove the largest arcs to obtain graph $G_{NRG} = (V\ E,\ _{NRG})$. E_{NRG} can be constructed using arcs *(u,v)* of the unitary graph as follows:

$$(u,v) \notin E_{NRG} \Leftrightarrow \exists w(d(u,w) < d(u,v) \wedge d(v,w) < d(u,v)) \qquad [1.2]$$

Thus, nodes *u* and *v* have a common neighbor, *w*, and so we can remove arc *(u,v)*.

1.2.2. *Connectivity*

DEFINITION 1.1.– *A graph G = (V E,) is connected if and only if, for every pair of distinct peaks, a chain exists that links them.*

The concept of a path is necessary to define connectivity.

DEFINITION 1.2.– *A path is a series of nodes with the form:* $s = u, a_1, \ldots, a_n, v \,|\, u, v, a_i \in V$ *, where u is the message source and v is the destination.*

DEFINITION 1.3.– *A graph* $G=(V,E)$ *is* λ*-connected if and only if* $\forall u,v\in V,\exists s|P(s)\geq\lambda$, *where P(s) is the probability that v will receive the message sent by u over path s.*

Two different methods can be used to calculate $P(s)$:

– Either we consider that v can receive the message from u via any node on path s:

$$P(s)=P((a_1,v)\cup\ldots\cup(a_n,v))=\sum_{k=1}^{n}-1^{k+1}\sum_{1\leq i1<\ldots<ik\leq n}P((a_{i1},v)\cap\ldots\cap(a_{ik},v)) \quad [1.3]$$

– Or we estimate that the message can only be received at v via a_n, giving us:

$$P(s)=P(a_1,a_2)\cap\ldots\cap P(a_{n-1},a_n)=\prod_{i=0}^{n-1}P(a_i,a_{i+1}) \quad [1.4]$$

1.2.3. *Mobility*

A node in an *ad hoc* network is able to move, and it can join or leave the network at any time, which causes the appearance or disappearance, respectively, of links. The movement of nodes is generally random; highly developed routing protocols are needed to control node movement. On the one hand, node movement is an important factor in the provision of major *ad hoc* network services; on the other hand, however, it causes difficulties with regard to routing function, network connectivity and energy optimization. Moreover, in a mobile environment, routing protocols require that a route has been completely mapped before any possible data transmission. It means that a transmitter node may have to wait a long time for the new path to be found. Simulations in [CAR 99] on routing protocols in an *ad hoc* network demonstrate the impact of mobility on the transmission of packets. As mobility increases, the number of packets transmitted drops. The additional cost in terms of quantity of control information exchanged increases, thus reducing the bandwidth available for data transfer. Some protocols are more sensitive than others to mobility; however, the fact that a node moves does not necessarily cause major reductions in the performance if the graph's connectivity is not affected (or it is only slightly affected) when it is necessary to transmit packets. Modeling mobility is among the difficulties encountered during *ad hoc* network simulations [ROH 14], but several models currently exist that include node mobility, including Random Walk, referenced point group mobility (RPGM), Manhattan and Freeway.

1.2.4. *Networks: wireless mesh network (WMN), wireless sensor networks (WSN) and mobile ad hoc network (MANET)*

A node in an *ad hoc* network is able to move; it can join or leave the network at any time, which causes the appearance or disappearance, respectively, of links. The movement of nodes is generally random; highly-developed routing protocols are needed to control node movement. On the one hand, node movement is an important factor in the provision of major *ad hoc* network services; on the other hand, however, it causes difficulties with regard to routing function, network connectivity and energy optimization. Moreover, in a mobile environment, routing protocols require that a route be completely mapped before a possible data transmission. It means that a transmitter node may have to wait a long time for the new path to be found. Simulations in [CAR 99] on routing protocols in an *ad hoc* network demonstrate the impact of mobility on the transmission of packets. As the mobility increases, the number of packets transmitted drops. The additional cost in terms of quantity of control information exchanged increases, thus reducing the bandwidth available for data transfer. Some protocols are more sensitive than others to mobility; however, the fact that a node moves does not necessarily cause major reductions in the performance if the graph's connectivity is not affected (or it is only slightly affected) when it is necessary to transmit packets. Modeling mobility is among the difficulties encountered during *ad hoc* network simulations [ROH 14], but several models currently exist that include node mobility, including Random Walk, RPGM, Manhattan and Freeway.

There are three types of *ad hoc* networks: wireless mesh networks (WMNs), wireless sensor networks (WSNs) and mobile *ad hoc* networks (MANETs). These networks have several similarities, as well as certain differences that logically involve different solutions.

1.2.4.1. *Wireless Mesh Network*

The mesh network is an emerging new technology that constitutes a particular case of an *ad hoc* network. It combines the benefits of *ad hoc* networks previously detailed with the benefits of terrestrial networks, notably in terms of available output due to the organization of the network with reduced-mobility nodes devoted to routing and mobile client nodes. WMN networks are composed of two essential elements: mesh routers and mesh clients. Mesh routers are links, while the mesh client connects to the closest mesh router and uses the *ad hoc* infrastructure to access the services of the *ad hoc* network. The difference between a classic *ad hoc* network and a mesh network is the restriction of routing functionalities to a subgroup of the network formed of mesh routers. WMNs diversify the capacities of an *ad hoc* network by introducing a hierarchy into the network [AKY 05].

1.2.4.2. *WSN sensor network*

A sensor network is an *ad hoc* network that is composed of nodes equipped with control units and measurement units. These units are characterized by a reduced processing and storage capacity due to their miniature sizes (on the order of 1 cm^3). The sensor nodes periodically send their acquired data to a special node called a sink node, which is responsible not only for the collection of reports but for the broadcasting of requests for the types of data required to the sensors via request messages [AKY 02, PIL 14, KAR 14]. The rapid deployment, reduced cost, self-organization and breakdown tolerance of WSN networks are characteristics conducive to their use in various domains, including military (chemical radiation, battlefield analysis), environment (temperature, humidity, seismic activity), medical (internal body imaging) and security (intrusion, surveillance, heating, etc.).

1.2.4.3. *MANET mobile network*

MANETs are characterized by a strong node dynamic with topology that is highly variable due to the frequent changes in node positions. To reach its destination, a message passes through several relay nodes. Unlike WSNs and WMNs, the nodes in MANETs are all mobile, and communication can take place between any of the nodes in a network. It means that the failure of any node is significant and must be handled quickly by specialized algorithms. Routing, topology control, and self-organization are basic techniques for the functioning of MANETs.

1.2.5. *Routing*

Routing is a function that consists of determining the route of each packet from a known source toward one or more destinations. Routing can also be defined as the task of transporting data from source nodes to destination nodes [XIA 08, TOU 99]. If a single destination is involved in the communication, this task is known as unicast routing, but when all the nodes in the network, or just a group of nodes, are receiving data, then we speak of broadcast and multicast routing, respectively [TAV 06]. The objective of routing algorithms is to find the shortest path between nodes, either in terms of number of hops, or in terms of link length (time). The algorithm functions in terms of time when links are dropped or re-established, or during a change in traffic conditions within the network. There are networks in which the packet can use different routes (this is generally the case with the Internet networks); these types of networks are called packet-switched networks and are always equipped with dynamic routing control functions. On the other hand, in networks based on packet switching architecture (such as traditional telephone networks and asynchronous transfer mode (ATM) networks), the routing decision is made on each connection, and all connection packets use the same path. The routing will be more complicated, and more complex problems may appear, when the network's users are mobile, and in a more complicated manner when the nodes are mobile themselves; this is the case with the satellite networks that use satellites in low orbit, called LEO, and *ad hoc* networks as well.

In *ad hoc* networks, decentralized routing is already used; therefore, the nodes will be more sensitive to changes in network topology. The instability of the wireless communication medium, the limitations of energy and bandwidth and the mobility of nodes introduce more difficulty and complexity during the design of routing protocols for mobile *ad hoc* networks. In MANETs, depending on the manner of establishing and maintaining routes during the transport of data from mobile nodes, we distinguish three principal categories of routing protocols: proactive, reactive and hybrid.

1.2.5.1. *Proactive protocols*

Proactive protocols are based on classic link-state and distance-vector algorithms. Each node holds routing information that concerns all the nodes in the network. This information is stored in routing tables that are updated with each topological change in the mobile *ad hoc* network in order to reconstruct the routes. Among the most widely used proactive routing protocols are: destination-sequenced distance vector (DSDV), Wireless Routing Protocol (WRP), global state routing (GSR) and optimized link state routing (OLSR) [COR 99, CAR 03].

1.2.5.2. *Reactive protocols*

Reactive routing protocols are also called on-demand protocols. They create and maintain routes according to the communications' needs in the network. When a transmitter node has need of a route, it launches a route discovery procedure [MAR 00]. Reactive protocols can be divided into two subcategories: source routing and hop-by-hop routing. The advantage of reactive protocols is that they offer greater adaptability to the topological changes of an *ad hoc* network [BAD 03, CHI 05] following the use of very recent "fresh" data for routing. Dynamic source routing (DSR), *ad hoc* on-demand distance vector (AODV) and core extraction distributed *ad hoc* routing (CEDAR) are the routing protocols that have been used with great frequency in recent years [PER 03].

1.2.5.3. *Hybrid protocols*

This category combines the first two types of routing protocols to achieve a shorter response time by taking advantage of the benefits of proactive and reactive protocols. In a hybrid protocol, the network is broken down into small zones where routing inside each zone is ensured by the proactive protocol, while routing between the different zones is based on the reactive protocol. Zone Routing Protocol (ZRP) and zone-based hierarchical link state (ZHLS) are among the best-known hybrid protocols [TOU 99].

1.2.6. *Weak security*

Ad hoc networks are notable for their weak security against various attacks, whether internal or external. An *ad hoc* network can be attacked in its basic functions such as routing, which is vital for the network to function properly. Attacks such as black holes, identity spoofing and wormholes disrupt routing protocols via multiple tactics to prevent them from functioning correctly. The Sybille attack is effective against routing algorithms, data aggregation, the equitable distribution of resources and the detection of malevolent nodes. Sybille has aroused particular interest in the scientific community, as much because of its originality as of the difficulty in finding countermeasures at a reasonable cost in *ad hoc* networks [DOU 02]. Security issues in *ad hoc* networks are therefore quite complicated, as we seek to authorize new nodes to participate in the network while avoiding nodes that will reroute or disrupt the routing function [ZHU 14, NOV 14].

1.2.7. *Access to the environment*

The major challenge in *ad hoc* networks consists of knowing who has permission to transmit at a given time, which needs the design of protocols to manage this type of situation. Medium access control (MAC) protocols are expected to accomplish this task. The MAC layer contains random access protocols that are generally characterized by low output due to several factors that can influence the network's quality of service (QoS). Among the limitations faced by these MAC protocols during their operation are collisions, successive retransmissions, delivery times, error rates, etc. In addition, the power supply to mobile nodes in MANETs is dependent on power sources of limited capacity. Thus, in order to ensure the functioning of

MANETs for a sufficient period of time, it is necessary to apply techniques to optimize and/or save energy by considering all the sources of energy overconsumption or waste adopted by MAC access protocols. MAC access in *ad hoc* networks, and in wireless networks generally, has been defined by the IEEE work group according to standard 802.11 [MÜH 02, HAR 04]. The main objective of the MAC layer according to this standard consists of providing reliable data services for upper layer protocols.

1.3. MAC protocols in mobile *ad hoc* networks

Currently, mobile *ad hoc* networks use the distributed coordination function (DCF) protocol or its improvement, the enhanced DCF (EDCF) protocol, which is structured on the carrier sense multiple access with collision avoidance (CSMA/CA), when accessing the environment. Random access methods are generally grouped into two main families: ALOHA and its derivatives, and CSMA and its derivatives [BEN 07]. In the next sections, we will introduce the classic random access methods in order to set down the basic concepts of the new environmental access techniques [GAJ 14, MUK 14, MOC 12, DUA 14, SEN 14, JAC 14, MOK 14].

1.3.1. *ALOHA*

The name of this method comes from experiments conducted at the University of Hawaii to link computer centers scattered across several islands. In it, nodes transmit packets unconditionally as soon as they receive them. There is no support listening before transmission. In addition, the propagation time of signals on the satellite channel is a limiting factor, since nodes are warned of a collision only 20 ms after data transmission. In the event that a data transmission has not been executed correctly, the node will retransmit the packets after a random period of time. This access method therefore has a low satellite channel use rate, approaching 20%; techniques exist that are similar but have been modified to achieve better performances [ALT 87]. Transmission in this protocol is completely decentralized. The basic principle is as follows:

If you have a message to transmit, transmit it.

If the message interferes with other transmissions, try to send it later.

The probability that *n* packets will arrive in two different packet times is given by:

$$P(n) = \frac{(2\lambda)^n e^{-2\lambda}}{n!} \qquad [1.5]$$

where λ is the traffic load.

The probability $P(0)$ that a packet will be successfully received without collision is:

$$P(0) = e^{-2\lambda} \qquad [1.6]$$

So, the output Th is given as follows:

$$Th = \lambda.P(0) = \lambda.e^{-2\lambda} \qquad [1.7]$$

Hence, the maximum output value of this technique:

$$Th_{\max} = \frac{1}{2.e} \approx 0.184 \qquad [1.8]$$

1.3.1.1. *Slotted ALOHA (SALOHA)*

An improvement has been added to the original ALOHA protocol is called ALOHA with temporal segmentation. This protocol introduces user synchronization. Time is cut into fixed-duration intervals called slots. Users cannot start transmission until the beginning of each slot. The probability of collision is thus reduced, and the maximum output is doubled. Thus, if there is not another user in transmission mode at the start of the time slot (TS), the probability function Th_i will be equal to:

$$Th_i = \frac{P_i}{(1-P_i)} \prod_{i=1}^{n} (1-P_i) \qquad [1.9]$$

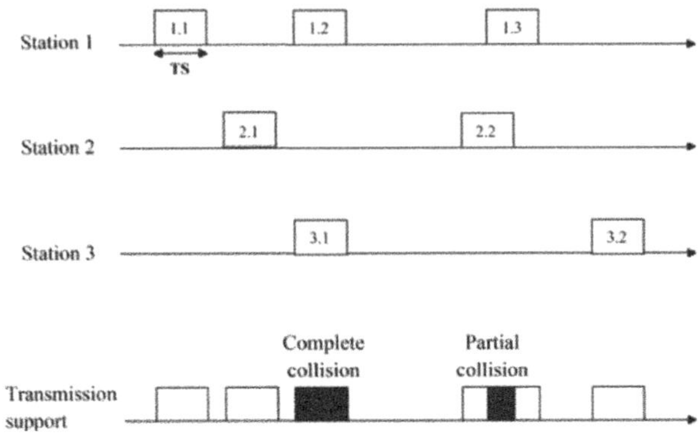

Figure 1.1. *Collision in the ALOHA protocol*

If Th is the traffic output and λ is the traffic, we can write:

$$Th_i = \frac{Th}{n} \text{ and } P_i = \frac{\lambda}{n}$$

$$\frac{Th}{n} = \frac{\lambda}{n} \cdot \left[\frac{1}{(1 - \frac{\lambda}{n})} \right] \cdot \prod_{i=1}^{n} (1 - \frac{\lambda}{n}) \qquad [1.10]$$

which means that the maximum output of SALOHA is: $Th_{\max} = \frac{1}{e} \approx 0.37 \text{ (Paq / TS)}$

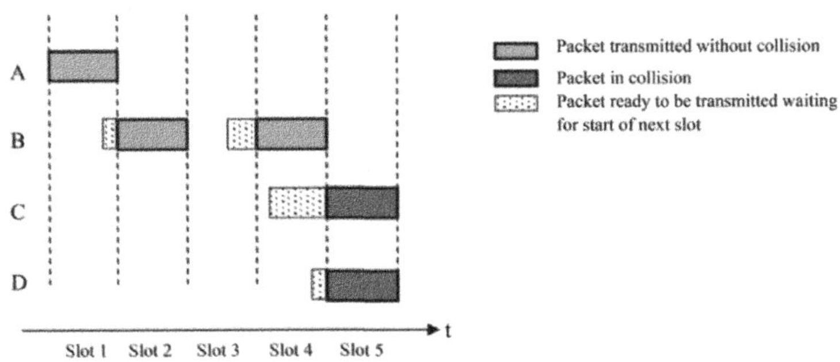

Figure 1.2. *Avoidance of collisions in SALOHA. For a color version of the figure, see www.iste.co.uk/benslama/adhocnetworks.zip*

1.3.1.2. *Multi-copy ALOHA*

When *m* copies of a packet are sent in the SALOHA (multi-copy) technique, the probability of successful transmission for this packet, or the probably that one packet of the *m* copies sent, will not enter into a collision will be higher than when a single copy is sent (S-ALOHA). This is true only when the other packets are sent in a single copy or when traffic in the channel remains stable without disruptions. In order to maximize the probability of successful transmission, we will assume that all users are transmitting the same number of copies (*m*) [MUH 04]. Multi-copy ALOHA is generally designed for satellite systems offering a higher probability of successful transmission, multichannel ALOHA systems or ALOHA systems with reservations, with a very high probability of success.

To assess output in this algorithm, we consider that the arrival of packets is a Poisson process. For simple SALOHA, we assume that the average retransmission period is larger than five slots, while the average period value for the *m* copies including the first transmission must be as large as five slots. So:

$$\lambda = \lambda_1 + \lambda_2 + \lambda_3 + ...\lambda_n = \sum_{i=1}^{n} \lambda_i \qquad [1.11]$$

The average number of copies per packet is:

$$N = \lambda^{-1}.\sum_{i=1}^{n} i.\lambda_i \qquad [1.12]$$

The probability P_i that packet *i* will be successfully received is:

$$P_i = 1 - prob[all\ copies\ are\ in\ collision] = 1 - \left(1 - e^{-N\lambda}\right)^i \qquad [1.13]$$

So, for *K* copies, we have:

$$Th_k = \lambda.P_k = \lambda\left[1-(1-e^{-k.\lambda})^k\right] \qquad [1.14]$$

To maximize the probability of successful transmission, a single copy must be transmitted when traffic in the channel is greater than 0.48, and when $\lambda \in [0.28, 0.48]$, we can only transmit two copies. The same figure shows that when λ is small, the probability P will reach the maximum value ($P = 1$) using a larger number of copies.

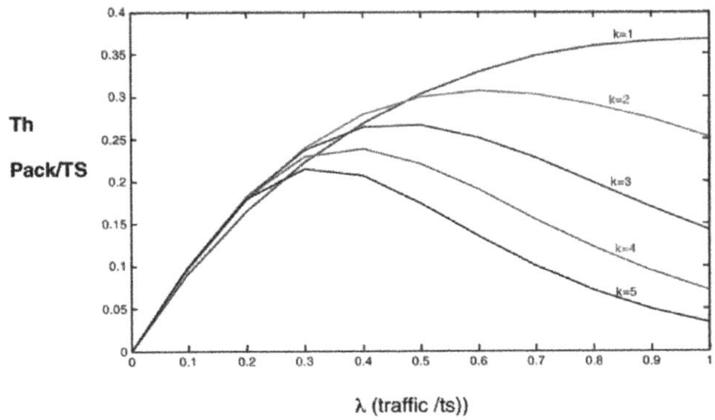

Figure 1.3. *Output in multi-copy ALOHA. For a color version of the figure, see www.iste.co.uk/benslama/adhocnetworks.zip*

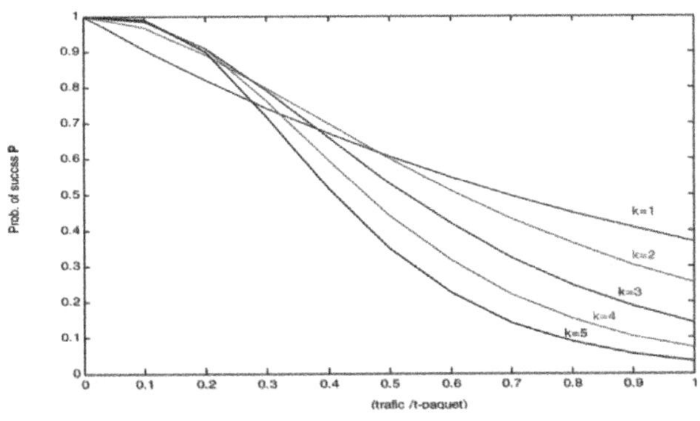

Figure 1.4. *Probability of success in multi-copy ALOHA. For a color version of the figure, see www.iste.co.uk/benslama/adhocnetworks.zip*

Consequently, random access methods such as ALOHA, SALOHA and multi-copy ALOHA have relatively modest performances and a loss rate that is too high for satellite or *ad hoc* contexts. In fact, communication systems in satellite networks cannot function using access methods with a high collision rate, as retransmission necessarily introduces excessive delays. Therefore, the use of these access methods is limited to the conveyance of signals, identification messages or small control and acknowledgment packets.

1.3.2. *CSMA*

In CSMA, mobile devices transmit only when the channel is free in order to avoid collisions. The principle of CSMA can be explained as follows: a node wishing to transmit in a channel first listens to the communication environment (mesh with bus). If the environment is free it transmits; if not, it waits for a specific amount of time. If the transmitter has not received the information after a given time, it supposes that a collision has taken place. After collision, the node waits for a random period and then retransmits [UYA 14]. There are several variants of CSMA, each of which possesses different behavior in the event of an occupied environment [LES 12, YAN 13]; thus, we distinguish between non-persistent CSMA, persistent CSMA and P-persistent CSMA. In the case of non-persistent CSMA, if the channel is occupied, it waits for a random amount of time and then transmits, while in the case of persistent CSMA, when the channel is occupied, the node continues listening to the channel until it becomes free, and then begins the transmission. In the case of P-persistent CSMA, the node acts differently; when the channel is free, the node transmits with probability P and waits for a period with probability $(1-P)$, but when it finds the channel occupied, it continues listening to the channel until it becomes free. If the transmission has taken a long time, the node begins listening to the transmission channel again.

1.3.2.1. *CSMA with collision detection (CSMA/CD)*

In CSMA/CD, before any attempt at transmission, the node makes sure that the channel is not already being used (carrier detection), and when the channel is free, the node rechecks the channel after a random period; if the channel is still free, the node sends its packets. However, this does not confirm that the packets have been successfully received. In reality, one or more nodes may send their packets simultaneously following this procedure, causing a collision, in particular when the network is crowded. This procedure is currently used mostly in wired networks, where collision detection is based on the type of electromagnetic propagation on a cable.

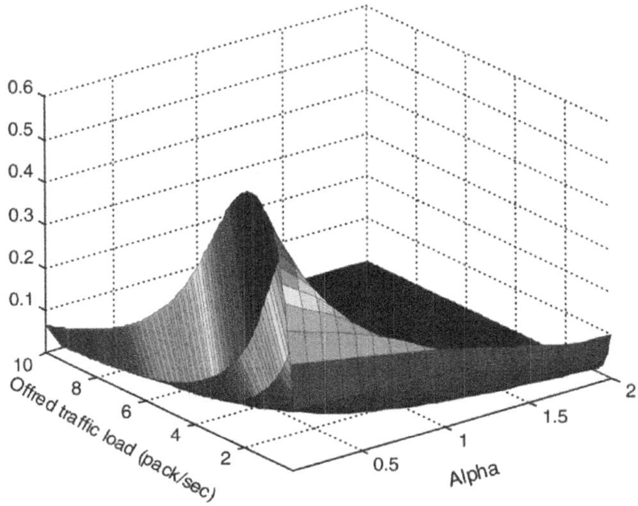

Figure 1.5. *Non-persistent CSMA output in 3D in relation to* λ *and* α*. Output is maximized when* $\alpha \in [0.1, 0.5]$ *and* $\lambda \in [1.6, 2.5]$*. For a color version of the figure, see www.iste.co.uk/benslama/adhocnetworks.zip*

1.3.2.2. *Standard 802.11 and the DCF algorithm*

The IEEE 802.11 standard was created in 1997 by the IEEE group. It describes the physical (PHY) layer and the data link layer, which is divided on two sublayers: the medium access control (MAC) layer and the logical link control (LLC) layer [RAM 14] of wireless networks characterized by a theoretical output of between 1 and 54 Mbit/sec, and a range that varies from 1 to 100 m depending on velocity and protocols used. Initially, frequency hopping spread spectrum (FHSS), direct sequence spread spectrum (DSSS), infrared (IR) and orthogonal frequency division multiplexing (OFDM) standards were adapted to the PHY layer, in which each of them used a different technique to spread the spectrum. The major problem with these standards was the lack of compatibility between them. However, many improvements have been made to these protocols, with new versions appearing to eliminate the drawbacks of the old versions and add benefits for the proper functioning of wireless networks, particularly impacting output and QoS. Three new layers have been designed for the PHY layer: 802.11b (wireless fidelity (WiFi)), 802.11a (WiFi5) and 802.11g, with the latter modified by 802.11n [GIB 92]. Figure 1.6 illustrates the architecture of the PHY layer and the MAC layer according to standard 802.11 compared to that of the open systems interconnection (OSI) standard. The data link layer is subdivided into two sublayers: LLC sublayer and MAC sublayer. The Data Link layer of standard 802.11 uses the DCF algorithm for non-centralized wireless networks, a category to which *ad hoc* networks belong. The DCF can be used by all nodes and offers equitable access to the radio channel without any centralization of access management in a fully distributed mode.

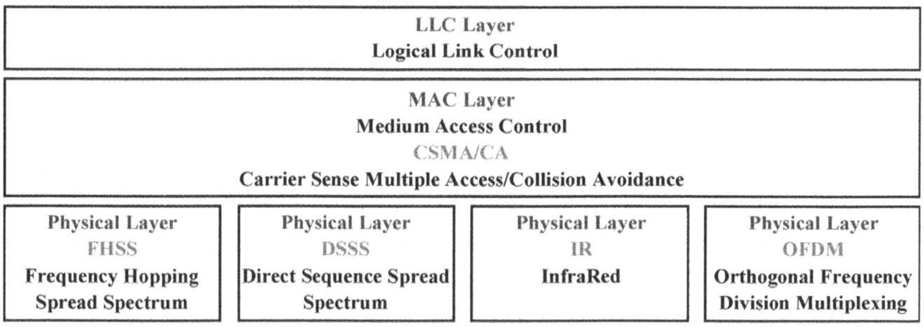

Figure 1.6. *PHY layer and data link layer in IEEE 802.11 [HAR 99]*

The first characteristic of the MAC layer consists of using acknowledgments to detect collisions and allow the retransmission of lost packets. In standard 802.11, the node can send in unicast mode toward a specific destination, or in broadcast mode toward multiple destinations. In the second case, there is no acknowledgment, and packets may be lost in collisions. The 802.11 standard in the MAC layer is based mainly on the use of Acknowledgment (ACK) frames in addition to request to send (RTS) and clear to send (CTS) information signals, which reduce errors due to interference and collisions on the radio channel in order to guarantee data integrity in the data link layer.

1.3.2.3. *CSMA/CA*

1.3.2.3.1. ACK principle

A node wishing to transmit listens to the channel; if it is occupied, the transmission is postponed. If the channel is free for a fixed period of time called distributed inter-frame space (DIFS), the node starts transmission after a random amount of time called backoff. The receiving node waits for the cyclic redundancy check (CRC) of the packet received and sends back an acknowledgment of receipt (ACK). Reception of the ACK indicates to the transmitter that no collision has taken place. If the transmitter does not receive an ACK, it retransmits the fragment until it obtains one, or abandons the attempt after a certain number of retransmissions.

To monitor network activity, the MAC sublayer works in collaboration with the PHY layer, which uses the clear channel detection (CCA) algorithm to assess the availability of the channel. To find out whether the channel is free, the PHY layer measures the power of the signal received by the antenna, called a received signal strength indicator (RSSI). The PHY layer determines if the channel is free by comparing the RSSI value with a predefined threshold, and then transmits a free-channel indicator to the MAC layer. In the opposite case, the transmission is postponed.

1.3.2.3.2. Definitions of IFS

According to standard 802.11, we differentiate four types of IFS: short IFS (SIFS), point coordination function (PCF) IFS (PIFS), DCF IFS (DIFS) and extended IFS (EIFS), listed in decreasing order of duration as follows:

– *SIFS* is the shortest distance between the four types of IFS. It separates a data packet from its acknowledgment. It is used for the transmission of ACK and CTS frames, responses to polling and barrages of frames sent by a single node. SIFS value is set by the PHY layer and is calculated so that the transmitting node is capable of switching in reception mode in order to decode the incoming packet.

– *PIFS* is used in PCF mode (Ethernet) and allows PCF transmissions to access the medium via the use of a smaller IFS than the one used for the transmission of frames in DCF. *PIFS* = *SIFS* + *TS*. *TS* = minimum time required to determine the state of the channel + round-trip time + propagation time. It corresponds to the minimum interval between two PHY carrier-detection operations. This value is dependent on the characteristics of the PHY layer being considered and is a constant specified by the standard for a given PHY layer.

– *DIFS* is most frequently used with SIFS. It is used in DCF mode as the minimum wait time before transmission. *DIFS* = *SIFS* + *2*TS*.

– *EIFS* is the longest of the IFS. It is used when a collision is detected and inhibits subsequent collisions [RAM 14].

IFS is used to define the degrees of priority of frames. When several nodes wish to transmit at the same time, the node with the highest-priority frames (acknowledgments, for example) begins transmitting first. This role will fall next to the node judged to have the highest priority, carrying frames related to network administration or traffic, which have time constraints. Finally, the least important frames concerning asynchronous traffic will be transmitted after a longer waiting period.

1.3.2.3.3. Backoff

When a node wishes to transmit its data packets, it listens to the channel for a DIFS; if the channel is free, the node begins sending its data. It can happen that two nodes detect the channel that is free at the same time after the DIFS and begin their transmissions at the same time, which leads to a collision. To reduce the probability of such a collision, after the DIFS, the node waits for a random amount of time called backoff time (BT), which is composed of a number of TS included between 0 and contention window (CW). After the DIFS, the node decrements its backoff by a TS step until it reaches the value of zero. If the node detects that the channel is occupied during BT, it starts over and waits for the channel to become free during the DIFS;

this time, if the channel is still free when the backoff reaches a value of zero, the node starts its transmission immediately. After each successful transmission, the receiving node sends an acknowledgment frame to the transmitter after an SIFS. However, if the transmission is unsuccessful, the node enters a double-length backoff [0 – 2*CW] in order to reduce the number of collisions in the event that multiple nodes wish to access the transmission channel. Here, note that not all the nodes that have postponed their access to the channel during the BT must start a new backoff, but they continue to decrement their last backoff counter in order to have priority during subsequent retransmission attempts.

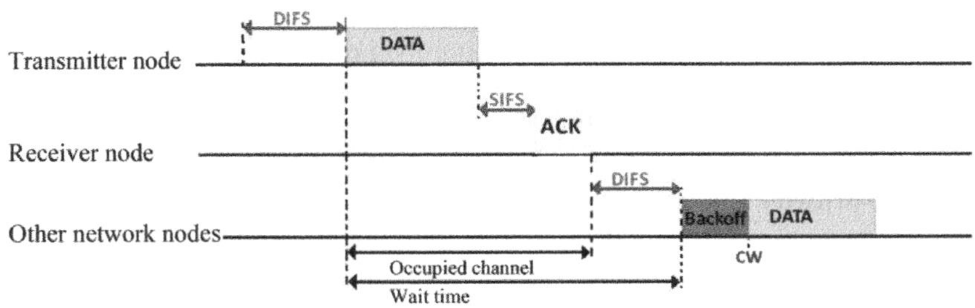

Figure 1.7. *IFS and backoff in CSMA/CA mode*

The size of the CW is initialized at a minimum value CWmin, which increases exponentially by doubling its size with each transmission failure. The CW size is reinitialized at CWmin after successful transmission or when it reaches the CWmax value after a limited number of retransmissions called retry limit (RL).

After a successful transmission, the transmitter node must conduct a backoff called post-backoff in order to separate successive frames.

Despite the integration of various techniques – backoff, IFS and ACK – the CSMA/CA protocol still experiences collisions. Hidden nodes and long frames are considered to be major sources of collisions. To limit collisions due to long frames and their multiple transmissions, data frames can be divided into fragments that can be transmitted sequentially as individual data frames. The advantage of fragmentation is that in the event of transmission failure, the error is detected sooner and there is less data to retransmit. The disadvantage is the overload introduced by the increase in the CW with the addition of additional acknowledgment frames. To solve the problem of hidden nodes associated with CSMA/CA, DCF defines a control mechanism called RTS/CTS.

1.3.2.3.4. RTS/CTS

This mechanism is also called virtual carrier listening. It is preventive against problems of hidden nodes and long frames, which introduce retransmissions that are costly in terms of time and spectral resources.

A node wishing to transmit first sends a small RTS control packet to request authorization from its receiver to begin data transmission. The RTS contains information about the source, destination and duration necessary to complete transmission. The RTS is received by the receiving node and by the other nodes in the network. The receiving node then responds by sending a CTS control packet that contains the same information as the RTS to inform the transmitter that the channel is free and it can begin data transmission immediately. At this time, the other nodes in the network, which have received at least one of the two control packets (RTS/CTS), each send a virtual network allocation vector (NAV) carrier listening indicator for a certain amount of time, delaying their transmissions until the expiration of the NAV's timer. If a hidden node has not received the RTS, it can then receive the CTS in order to update its NAV, which can minimize the problem of hidden nodes. NAV updating is ensured by the reception of RTS or CTS signals. The NAV informs each of the nodes not involved in the transmission of the length of time during which the channel will have to be occupied. At this time, the nodes may switch to power-saving mode (PSM) for this period of time in order to conserve energy.

An RTS frame is of 20 bytes and a CTS frame is of 14 bytes in size; these are therefore short frames with a low probability of collision. However, in the case of long frames, the collision rate is high due to multiple transmissions. To solve this problem, data frames can be divided into fragments that can then be transmitted sequentially and individually. The advantage of fragmentation is that, in the event of transmission failure, the error is detected sooner and there is less data to retransmit. However, an overload will be introduced by the increase in contention and also by the addition of more acknowledgment frames.

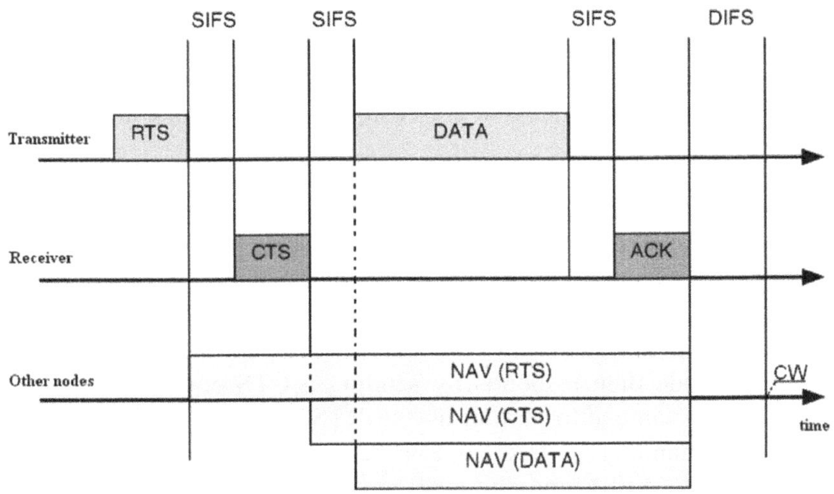

Figure 1.8. *NAV update*

In conditions like this, where there are significant overloads and inefficient use of bandwidth, an initiation threshold is implemented to limit the use of radio support. If the length of the data to be transmitted is lower than this threshold, transmission will take place without RTS/CTS; otherwise, the RTS/CTS mechanism is used.

It should be mentioned here that in broadcast mode, the RTS/CTS mode is inoperative. In this case, the frame is broadcasted toward all nodes, which means that there are multiple recipients and thus we can have multiple CTS simultaneously, which causes more collisions.

1.3.2.3.5. MAC frames in 802.11

802.11 frames are generally formed of four main parts: preamble, header, data and CRC. Each of these parts is characterized by a limited number of bits and a function that is well determined during the transmission of frames in an *ad hoc* network. The figure below explains the components of a MAC 802.11 frame.

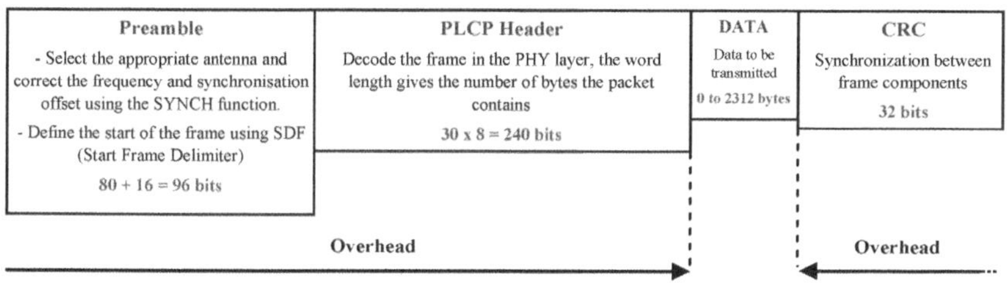

Figure 1.9. *Usual format of an 802.11 frame*

A physical layer convergence protocol–protocol data unit (PLCP-PDU) frame is composed of a PLCP header and of data from the MAC layer. The PLCP-PDU header contains two fields: preamble and header. Two types of preamble are defined: long one (192 bits) and short one (132 bits). A long preamble secures the connection to the network and thus the transmission. The word length of the PLCP_PDU gives the number of bytes the packet contains, which helps the PHY layer to correctly detect the end of the packet. In addition, the error checking header field is the CRC error detection field formed in 16 bits. MAC frames are sent with outputs ranging from 1 to 2, 5.5 or 11 Mbits with regard to the 802.11b [SHA 14, BAI 12].

– *ACK frame*

RA: Address of the receiving node (receiver address). This is the address indicated in the *Address 2* field of the frame preceding the ACK frame.

Duration is equal to zero or the value of the preceding *Duration* field minus the time requested to transmit the ACK frame and SIFS interval.

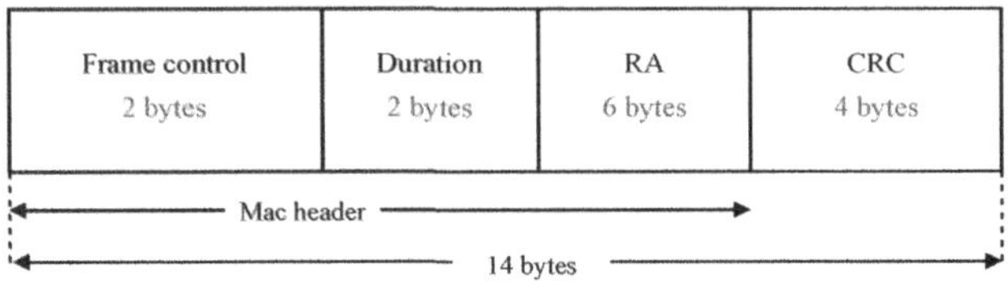

Figure 1.10. *ACK frame*

– *RTS frame*

TA: Address of transmitter node (transmitter address).

Duration is equal to the time necessary for the transmission of the management frame or of subsequent data, plus a CTS frame, plus an ACK frame and plus three SIFS intervals.

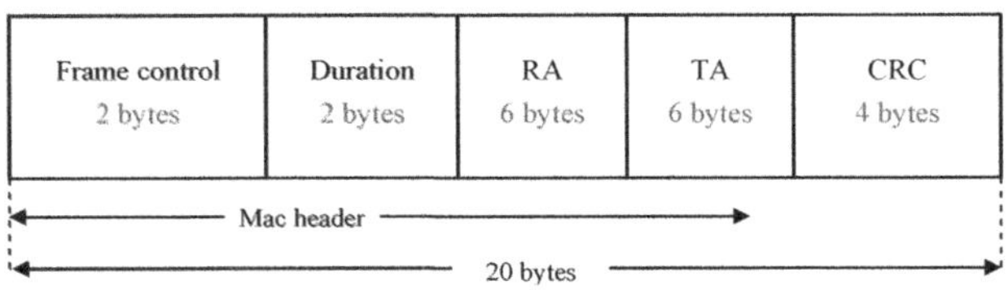

Figure 1.11. *RTS frame*

– *CTS frame*

RA: Address of the receiver of the CTS frame, directly copied from the TA field of the RTS.

Duration is equal to the RTS duration minus the transmission time of the CTS frame and one SIFS interval.

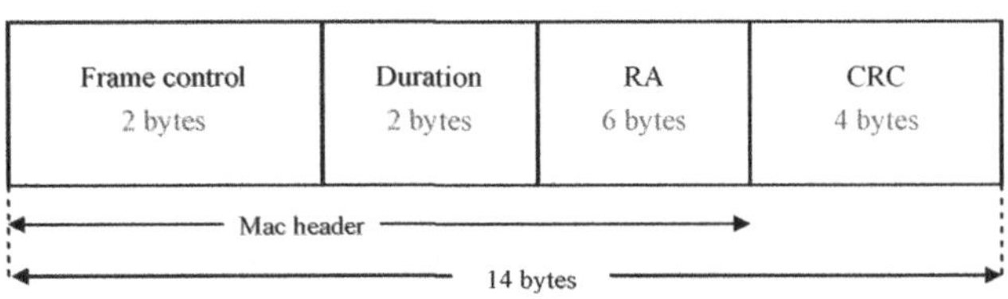

Figure 1.12. *CTS frame*

The frames used in standard 802.11 [ZHA 13a, PAV 14, LEE 14, SZO 14, SOR 14, SWA 14] usually follow the format shown in Figure 1.9; they are divided into three main types, specifically:

- *Data* frames for data transmission,

- *Control* frames to control access to the support (for example ACK, RTS and CTS),

- *Management* frames to exchange management information but without transmission to the upper layers.

At the time of transmission, a MAC header and a PHY header are added to the data generated by the upper layers. The transmission velocities of some of these parts can vary. The PHY header is sent at a constant velocity of 1 Mbit/sec, while the MAC header can be sent at a velocity that may reach 12 Mbit/sec. Control and acknowledgment packets are sent at different velocities.

An example of a 1,032-byte frame sent at 11 Mbit/sec can be broken down as follows:

– A DIFS with 50 µsec, a backoff with 0 to 31 TS of length 20 usec (0 to 620 µsec),

– *Data* frame of 987.7 µsec with:

  - *PHY header* of 192 µsec at 1 Mbit/sec,

  - *MAC header* of 24.7 µsec at 11 Mbit/sec,

  - *Useful data* of 771 µsec at 11 Mbit/sec.

– A *SIFS* of 10 µsec and the 192-µsec *PHY header acknowledgment* at 1 Mbit/sec,

– *MAC header acknowledgment* of 56 µsec at 2 Mbit/sec.

Theoretically, the maximum possible outputs can be calculated according to the size of the packets used during transmission.

1.4. Energy consumption in *ad hoc* networks

Energy sources in *ad hoc* networks are provided mainly by cells or batteries, which power the nodes during their operations in the network. These batteries are of limited capacity and can cover node activity for a reduced period of time in terms of several parameters: number and type of operations carried out; types of transmitters and protocols used in the network; and network mobility distance. Nodes in *ad hoc* networks generally consume energy [FEE 01] when they transmit data toward a pre-determined destination (transmission) or when they receive data (reception), as well as when they listen to the channel and during hibernation. Without energy, the nodes cannot function, and in this event, the network remains inactive. Since the energy supplied to mobile nodes is practically limited by sources (batteries) with a short lifespan, the saving, control and optimization of this energy in *ad hoc* network are of vital importance and currently pos one of the greatest challenges facing scientific researchers.

Sources of energy consumption in *ad hoc* networks can be linked to two principal operations: consumption related to communications and consumption related to the processing and analysis of information. Communication practically requires the use of transmitters at the source, en route and upon arrival at the receiving node. The objective of the transmitter is to generate original packets, control the route and redirect packets toward another recipient node. The objective of the receiver is to receive data, control the packets received and transfer packets toward other destinations. The quantity of energy consumed varies therefore according to the type of transmitter used. An example is given in [MAK 07] of a Proxim RangeLAN2 2.4 GHz 1.6 Mbps Personal Computer Memory Card International Association (PCMCIA) card, which requires 1.5 W for transmission, 0.75 W for reception and 0.01 W in hibernation mode; thus, switching between transmission and reception for a node occurs in between 6 and 30 µsec. In addition, there is an example of the Lucent 15 dBm 2.4 GHz PCMCIA card, which consumes 1.82 W in transmission, 1.80 W in reception and 0.18 W in hibernation mode. Though not nil, power consumption in hibernation mode can be disregarded in comparison to other consumption values. Figure 1.13 shows the consumption of some transmission modules from the firm Ericsson, used to establish a high-frequency (HF) link in a global system for mobile communication (GSM) cell site.

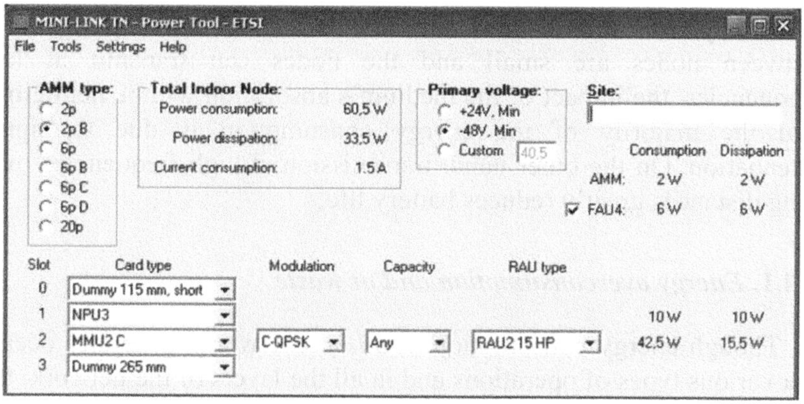

Figure 1.13. *Variation in energy consumption according to types of transmission modules*

Consumption related to the processing and analysis of data before transmission and after reception can generally be summed up in operations of calculation, sampling, modulation, encoding, decoding, filtering, compression, A/N conversion, etc. This is demonstrated in [MAK 07]: an msp430 micro-controller can consume between 14 and 23 mW depending on family, while the quasi-delay insensitive (QDI) 8-bit CISC MICA asynchronous microcontroller, which functions at 23.8 Mips (2.5 V), can consume up to 28 mW. Consequently, the more radio communications take place, the more the microcontroller must make calculations and the higher the proportion of energy consumed by it on the overall energy tally. The technique of data compression, which is used to reduce packet length, causes additional energy consumption due to the increased number of processing and calculation operations. Therefore, a relationship exists between the two sources of energy consumption, and protocols intended to diminish communications consumption may cause an increase in consumption related to data processing and analysis. In order to optimize energy consumption, we must control the balance between the two sources of this energy consumption.

An example of an underwater *ad hoc* network is given in [JUR 07]. The curve shows the influence of transmitter–receiver distance and frequency on the average lifespan of batteries. Battery life is constantly reduced with the increase in distance. When the distances between nodes are small and the nodes can transmit at low frequencies, the impact of the medium's absorption will be negligible, and the majority of the energy consumption is due to signal attenuation. On the other hand, transmission at high frequencies over long distances greatly reduces battery life.

1.4.1. *Energy overconsumption and/or waste*

Though energy consumption in *ad hoc* networks generally occurs via various types of operations and in all the layers of the network, the largest amount of consumption is due to communications. MAC protocols include sources of energy overconsumption and sometimes waste in their details, and these sources should be reduced or eliminated.

– Collisions in the MAC layer [KRA 98, YE 02] are responsible for retransmission phenomena that cause an overconsumption of energy, resulting in delays causing considerable losses in both energy and time (output). In reality, we cannot definitively eliminate retransmissions that occur because of errors made during transmission and during collisions. However, several solutions and techniques for reducing them are suggested in the literature [STE 97, JON 01, HAC 03, SIV 00, SIN 98, LI 01, SCH 01].

– We have seen that nodes in inactive or idle mode consume energy without carrying out any operations other than listening to the channel. This results in a loss of energy that can be minimized via practical techniques such as occasionally putting nodes into hibernation mode or turning off the transmitter for a pre-determined amount of time. The power aware multi access protocol with signaling (PAMAS), IEEE 802.11 PSM, sparse topology and energy management (STEM) and sensor-MAC (S-MAC) protocols, which are explained in [SIN 98, GAD 04, KAR 90], respectively, have developed this approach.

– In addition, large amounts of time and energy are gratuitously lost during the switching of mobile radios between Tx mode and Rx mode, and vice versa; this is the case with protocols based on slot-by-slot technology. An algorithm is suggested in [GAD 04] to avoid this energy loss, in which case reception and transmission slots are considered separately.

– Poor quality of the communication medium can cause many transmission errors and thus a high error rate, while the packets involved in these error transmission become useless, which means that the energy used during their transmission will be lost. In similar channel conditions, transmissions can be avoided until the channel is reestablished in order to avoid energy waste [ZOR 97]. Error control protocols based on automatic repeat request (ARQ) and forward error correction (FEC) techniques can also be used to conserve energy; these protocols are explaned in [LET 97, WOO 01, CAN 00].

– Most of these routing protocols rely on the number of hops as a metric in the choice of paths, while other protocols such as associativity-based routing (ABR) and signal-stability adaptive routing (SSA) rely mainly on link quality [JUB 87]. Unfortunately, these routing metrics have a negative impact on the lifespan of nodes and of the network in general, due to energy overconsumption for some nodes in favor of others.

– Choosing the wrong type of routing protocol can cause an undesirable overconsumption of energy. In [CHE 01], comparisons between several routing protocols based on experimental measurements have shown that the DSDV and temporally ordered routing algorithm (TORA) protocols increase energy consumption by 51.8% compared to the DSR and AODV protocols, which have a low energy consumption rate and greater stability in the network.

The lack of cohesion between MAC and routing protocols results in a risk of producing cutoffs and ruptures in network connectivity due to the inequitable use of nodes, as the routing protocol may call upon low-energy nodes that the MAC protocol has put into hibernation mode, and the MAC protocol may put high-energy nodes into hibernation mode that might be selected by the routing protocol to transport information. A solution is examined in [MES 11] that consists of establishing a sort of switching between the different protocols to achieve better energy conservation (up to 14%).

1.4.2. *Toward more efficient energy consumption*

Controlling energy consumption in *ad hoc* networks is of vital importance given the limited energy sources available, which are usually in the form of batteries or cells. During their activities, nodes consume energy in a non-uniform manner, and thus energy distribution in the network is inequitable. This is due to the quality of operating policies followed in the different layers of the network. Therefore, targeting efficient energy consumption calls on all of the layers in an *ad hoc* network. However, the MAC layer has the most influence on the energetic behavior of the network. We saw in the last section how MAC protocols can be responsible for energy overconsumption and even waste due to node behavior that is not well structured. Moreover, routing protocols play a large role in the loss or saving of energy, and cohesion between these two protocols leads to a balanced use of the nodes in *ad hoc* networks.

There are currently three axes of development for the optimization of energy consumption by communications in *ad hoc* networks:

– Energy savings in the case of the problem of energy loss in inactive mode. This is a matter of maximizing the duration of the nodes' hibernation mode.

– Control of transmission power, which consists of increasing network capacity and transporting data at minimal energy cost by allowing nodes to determine the minimum transmission power sufficient to maintain network connectivity.

– Load distribution, the principal objective of which is to balance energy consumption among mobile nodes.

In the next part of this chapter, we will examine the various solutions contributed for each network layer (according to IEEE standard 802.11) for efficient energy consumption in MANETs.

1.4.2.1. *Data link layer*

1.4.2.1.1. MAC sublayer

The MAC sublayer is responsible for providing reliability to the upper layers of the network in point-to-point connections established by the PHY layer. The objective of this layer is to minimize simultaneous accesses that cause collisions. MAC protocols are grouped into two major categories according to the method of access in each protocol; there are fixed and random protocols.

Energy management in the IEEE 802.11 protocol

The idea behind the 802.11 protocols is to wait for a random period of BT when the channel becomes free before starting transmission. The backoff mechanism limits the risk of collision but does not eliminate it completely, and when a collision occurs, a new backoff will be automatically initiated. However, on each consecutive collision, the size of the window will be doubled in order to reduce the chance that such collisions will occur again. This approach reduces the number of collisions, and thus conserves node energy; in addition, it introduces additional delays and a considerable lowering of output in the *ad hoc* network.

PAMAS protocol

This protocol is a development of an older protocol, MACA [KAR 90], with the introduction of a separate channel designed for control messages (RTS, CTS and busy tone). It is intended to save energy in *ad hoc* networks. In PAMAS, before a node begins data transmission, it must send an RTS via the control channel and await the CTS response of the receiving node. If it does not receive the CTS, the node enters a backoff period. However, if it receives the CTS, the node transmits its packets via the data channel. At this time, the receiver node sends a busy tone message through the control channel to inform the nodes monitoring the control channel that the data channel is occupied. At that point, any nodes unable to transmit or receive are instructed to turn off their radio interfaces to save energy.

In PAMAS, the use of an independent control channel allows nodes to determine when and for how long the radios will remain turned off. A node must turn off its radio interface either when it has no data to transmit and is not concerned by a transmission from a neighboring node or when it has packets to transmit but a transmission is in the process of being executed by its neighboring nodes. Each node can determine the amount of time during which it must turn off its radio using the probe protocol, as explained in [SIN 98]. In summary, PAMAS functions based on the principle that when a node is free/empty (free medium), no energy will be consumed. Research [AKY 05] has shown that the PAMAS protocol reduces energy consumption by at least 50% in networks with large communication loads (0.5–3 packets/sec/node). Numerous routing algorithms have been suggested for *ad hoc* networks with the objective of saving energy; some energy-optimizing routing protocols are presented below.

Geographic adaptive fidelity (GAF) protocol

GAF is a routing protocol that uses the node localization technique in *ad hoc* networks via the use of GPS. This protocol consists of creating virtual grids based on the area involved in routing by dividing this area into small zones such that for two adjacent grids Gx and Gy, all the nodes in Gx can communicate with all the nodes in Gy in order to ensure permanent network connectivity. One condition is necessary in this protocol, which is that a single node must be active in each grid, and that the other nodes in the same grid must be in hibernation mode for a fixed period of time. This technique saves energy and confirms the fidelity of the network. However, in some environments where nodes are highly mobile, routing fidelity may be reduced if an active node leaves the grid, which can cause data loss.

Switched port analyzer (SPAN) protocol

In the SPAN protocol [GAD 04], routing is done with the aid of coordinators. The selection of a coordinator node is based on the energy level it possesses and the number of neighboring node pairs it can connect. Coordinators remain permanently active in order to ensure multi-hop routing in the network, while the other nodes remain in hibernation mode and verify periodically whether they should activate to become a coordinator, or not yet. Via this process, SPAN ensures energy savings for nodes in hibernation mode on the one hand, and equitable energy distribution via the tactic of alternating nodes to replace the coordinator on the other hand, which results in a considerable extension of the lifespan of the network. Experiments have shown that the SPAN protocol gives better results than the GAF protocol, though GAF is simpler to implement than SPAN.

1.4.2.1.2. LLC sublayer

This layer is responsible for the security of communications via controlling errors in transmission, data encryption/decryption and packet retransmission. There are two main families of techniques used for error control: ARC technique [ATE 08] and FEC technique. These techniques are known for non-optimized use of bandwidth and energy due to successive packet retransmissions and the time elapsed during error correction. Great care must be taken when using these techniques in a wireless connection, where the error rate is higher due to noise, fading and disconnections caused by node mobility. In addition, equilibrium in this layer must be maintained between the various characteristics of the network in order to improve output, security and energy efficiency; for example, the improvement of channel quality via an encoding system may reduce output through the addition of redundant bits in packet transmission. In addition, if transmission power is increased to avoid interference, the batteries are exhausted and the duration of operation of MANETs is reduced. For this reason, recent research has suggested alternative techniques and new protocols for error control with more moderate energy consumption.

Adaptive error control with ARQ

A new protocol is proposed in [HEC 05, RAZ 14] for error control with optimized energy consumption. This protocol is based on the following three principles:

– Avoiding persistence during data retransmission,

– controlling the number of retransmissions according to probability of success,

– avoiding transmissions in mediocre channel conditions.

In this protocol, ARQ functions normally until an error is detected in the control channel following the absence of ACK signals; at that point, the protocol enters probing mode, in which a probing packet is sent to each *t* slot. This packet contains only the input bit and not the data bits so that it consumes less energy. Probing mode continues until the reception of a correct ACK signal, indicated that the channel has been reestablished, and the protocol reverts to normal mode and continues the transmission from the cutoff point. The results obtained in [HAN 04] show that in a channel with fading, the developed ARQ protocol is better than the classic ARQ in terms of energy consumption and number of packets transmitted.

1.5. Conclusion

Every day, *ad hoc* networks demonstrate their importance in daily life via the benefits they bring to services in different areas of economic, social and cultural activity. However, medium access and energy control remain critical points that are preventing us from benefiting from the maximum capacity of these networks. In this chapter, we have introduced *ad hoc* networks and their different characteristics, and we have discussed in detail the medium access phenomenon and the major problems present in the MAC layer. As we have emphasized, output and QoS currently constitute the major challenges of medium access protocols experiencing saturations and limitations with regard to multimedia applications and real-time services.

The energy factor is of primary importance in the management and control of *ad hoc* networks, notably in the evaluation of routing and medium access protocols. In this vein, we explained in the last section of this chapter the impact of energy on the functioning of *ad hoc* networks and the relationship between the effectiveness of these protocols and energy consumption. In addition, we have introduced the various solutions contributed by the MAC and LLC communication protocols attempting to guarantee more efficient energy consumption in dynamic MANET systems, which require additional energy in order to cover node mobility.

2

Game Theory and Communication Networks

2.1. Introduction

In many situations in daily life, an actor's performance depends not only on his or her actions, but also on those taken by others. This strategic interdependence is the field of predilection on which game theory is based. In recent years, this theory has marked the development of numerous disciplines in economic science, management, operational research, engineering, political science, information technology and biology, to name only a few. Understanding game theory, then, has become essential for anyone interested in these disciplines [DRE 61, FUD 91, MYE 91, GIB 92, CAM 05, OSB 94, FUD 98, OSB 00].

When the actions of the decision-making elements in a defined environment interact in the course of their activities, and when each of them seeks to optimize its gains, the performances of this environment can be weakened gradually, thus causing individual degradation in each element. In this type of situation, where the action of one element can have an impact on the decisions of others, the application of game theory can be highly beneficial. Indeed, game theory offers a tool for the analysis of contentious situations, based on subjective ideas and probability calculations when the action of each active agent is dependent not only on environmental conditions, but also on what the other agents decide to do in order to optimize its usefulness. For example, in the case of a seller and a consumer in a market, the seller will act in such a way as to maximize his profit in the face of competitors, for example by offering the best sale price. The consumer, for his part, seeks to purchase the item he wants at the lowest price after bargaining with the seller. In this example, each element in the market acts on his own account according to the principle of economic rationality, which stipulates that each person seeks to make the best decisions for himself, as in the case of a chess player wishing to win a game, who will employ the means he thinks are the best in order to do this.

This is precisely the paradigm of game theory: the builder of the model assigns gain functions and strategies to the gamers, who watch what happens when they choose strategies to obtain maximum gain [KAC 12].

The publication of the book *Theory of Games and Economic Behavior* by John Von Neumann and Oskar Morgenstern in 1944 [VON 44] launched the introduction of game theory. The book offered practical definitions for the use of this theory and introduced the idea that conflict could be analyzed mathematically, providing the necessary terminology to do so. Nor can we overlook the works of Cournot, Zermelo and Emile Borel, who suggested the basic concepts of this theory. But, well before all of this, the first work on game theory was done by John Von Neumann [VON 28, VON 44], who demonstrated the min–max theorem, which played – and still plays – an important role in game theory. According to Nash, man must rise above his own individual interest and take into consideration the strategies of the other elements in a group. John Nash's work on game theory earned him a Nobel Prize in economics in 1994. Since then, this theory has experienced major advances, particularly in so-called cooperative and repeated non-cooperative games. The importance of the role played by game theory was further recognized by the awarding of the Nobel Prize in economics to Robert Aumann and Thomas Schelling in October 2005.

The applications of game theory are innumerable, but we would cite in particular the following books, which focus on the regularization of public traffic watermarking, biology, sociology, optimal airplane configuration, multimodal filtering in the tracking of radar targets and the distribution of electrical current [SU 07, PAT 03, XIA 14, AL 14, RUN 10, CHA 14b, SU 14, REN 11, KAL 10, ASI 12, ODE 10, GUO 11, WU 14, LIN 14, LOP 07, IGA 11, KOU 14, CHA 14a, DEX 14, CHU 14, TOS 09, BAR 14, BIN 85, XIA 07, MU 13].

This chapter focuses on game theory and its applications in telecommunications, and more specifically in wireless networks. We will begin by introducing this theory as mathematical tool that includes a set of rules and techniques necessary for analytical developments. To this end, our first task is to provide definitions of what a game, a player, utility and game strategy are. Next, we will distinguish the various types of games that can exist, by comparatively describing their forms of representation. In practical terms, game theory cannot be discussed without assuming the famous theorem of Nash's equilibrium. To do this, we will give a detailed definition of this theorem and explain its importance in game theory, in order to apply it in several areas of science and life. Finally, we will give some examples of famous games known by game theory.

In recent years, game theory has made its mark in a great deal of scientific work and also in various fields such as telecommunications, where it has provided researchers with a processing tool used to study phenomena and issues in a different way, leading to interesting results. This method of analysis is based mainly on a set of subjective probabilities that take into consideration all the actions and decisions of the agents participating in this research subject. Now, wireless local networks (WLAN), which are installed in the most important city sites, such as libraries, airports and hotels, offer easy, fast Internet access at low cost. This is also the case with mobile *ad hoc* networks (MANET), which play a very important role today due to the services they offer in special situations (natural disasters, surveillance, etc.). These types of networks are becoming more and more open to participants belonging to different firms and authorities. Nodes in the network tend to be independent in order to optimize their quality return. How can nodes act independently? What is the impact of this independence on network performance? What is the best way to deal with these types of situations in order to arrive at a solution that is agreeable to all the elements of the network? The answers to these questions may give rise to new ideas that will be useful for the design of new networks in the future. To illuminate these points in question, in the second part of this chapter we will discuss the relationship between game theory and telecommunications, and the benefit it may have for the various functions and services of telecom networks. Finally, we will end the chapter with some recent studies of applications of game theory in wireless networks.

2.2. Introductory concepts in game theory

2.2.1. *Game*

A game can be defined as a set of rules that controls the behavior of a group of active elements (agents), and which determines their gains on the basis of the decisions and actions they undertake. An agent has entered a game if his utility and gains are affected not only by the actions he undertakes, but also by the actions of the other agents participating in this game. The author of [EBE 13] defines a game as a situation in which individuals (players) are led to make choices from among a certain number of possible actions called strategies, where each strategy is a complete description of the way in which a player intends to play from the beginning to the end of the game, in a predefined context known as the rules of the game. The result of this game constitutes an outcome, with which a positive or negative gain is associated for each of the participants. The rules of a game sometimes specify the order in which agents make their choices, and designate the utility each agent can have in the face of the decisions he undertakes.

2.2.2. *Player*

Any person or gent who participates in a game and is capable of making a *decision* is called a player. A player can be an agent, a company, a government, a consumer, a subscriber, a node, etc. depending on the field of activity in which the game is taking place. In game theory, and according to the type of relationship between the players, games can be cooperative or non-cooperative. We can also distinguish between perfect and imperfect information games, depending on the level of information possessed by the player before he makes his decisions. Each player engaging in interactions within the game acts on his own account in order to maximize his interests according to the principle of rationality.

Depending on the number of players, there are two-player games and n-player games. In [FUD 91], the author indicates that it is sometimes useful to explicitly include agents called pseudo-players, who undertake random actions at predetermined points in the game with probabilities that are also predetermined. Formally, a player is designated by the index $i, i \in N$. Extension to the case of an infinite number of players does not pose any particular design problems. $N = \{1,...,n\}$ represents all of the players, and we assume that there is a finite number of players.

2.2.3. *Strategy (pure and mixed)*

The term "strategy" refers to the actions taken by a player in an interactive game. In other words, a strategy is the complete specification of a player's behavior in any situation. The concept of strategy is useful, since the actions a player wishes to choose often depend on the past actions of other players. In reality, it is quite rare to predict the actions of a player unconditionally, but we can often predict the way in which he will respond to external conditions. For now, keep in mind that a player's strategy represents a set of instructions available to him, which indicate to him the actions to select in every possible situation [KAC 12, RAS 01, GUE 10]. There are two types of strategy; pure strategy and mixed strategy. *Pure strategy* reflects an action or series of actions chosen by the player with certainty, while *mixed strategy* is defined as a distribution of probability over a set of pure strategies.

Formally, we use s_i to designate the strategy of player *i*, $i \in N$. So, s_i is the rule that tells player i which action to choose at each stage of the game. We use S_i to define the *set of strategies*, or strategy space composed of all the strategies available to a player. Therefore $S_i = \{s_i\}$, and $s = (s_1, ..., s_i, ..., s_n) \in S_1 \times ... S_i \times ... S_n \equiv S$ is the outcome of a game, $S = \prod_{i=1}^{n} S_i$, meaning a combination of strategies based on one strategy per player. We use $s_{-i} \in S_{-i}$ to designate all the strategies chosen except the strategy of player i.

2.2.4. *Utility*

Utility is the negative or positive gain that results from a player's actions. The objective of each player in a conflict is always to maximize his gain in relation to the conditions that surround him. A player's utility can depend not only on his decisions, but also on those of all the other players as well. In addition, depending on the type of game, utility can be manifested in multiple qualities; for example it may be a price in a market, a number of points in a game of chess, a time in a race, a success rate in a school, a salary in a company, the robustness of an industrial mechanism, the power of a physical phenomenon, the error rate of a digital transmission, etc. Finally, we say that a conflict game is in equilibrium when all of the players are satisfied with regard to their individual utilities acquired. In the literature, utility may be expressed by the words *gain, benefit, payoff*, etc.

Formally, in a game G we use $u_i(s) \in R$ to designate the utility function of player i. Thus we can say that the utility of player $i \in N$ depends not only on his strategy s_i, but also on the strategies of the other players, summed up by s_{-i}. Therefore we can write that, if player i strictly prefers outcome s to outcome s', then $u_i(s) > u_i(s')$. And, if $u_i(s) = u_i(s')$, we say that the player is indifferent to either of the two outcomes.

2.2.5. *General classification of games*

There are multiple classifications of games according to the following criteria:

Number of players, number of strategies, type of relationship between players, type of gain, form of gain functions, number of steps in the game, and state of information.

However, generally speaking, there are three types of games:

– cooperative and non-cooperative games;

– perfect and imperfect information games;

– normal-form and extensive-form games.

2.2.5.1. *Cooperative and non-cooperative games*

A game is called cooperative if players can make agreements with one another that give rise to limiting relationships. This is the case, for example, with nodes that agree on a very precise link to route their data while freeing the other available links. In this case, we say that players have formed a coalition. However, when players are not able to form coalitions, we say that a game is non-cooperative. In this type of game all the strategic options offered to players are predetermined, which is not the case in cooperative games. Players in non-cooperative games cannot come together in coalitions, but they can agree on this or that outcome of the game, provided that they do not develop a limiting agreement. Non-cooperative games can also be in normal (strategic) or extensive form.

For informational purposes, the principal books addressing these different aspects of cooperation and non-cooperation are [JOR 09, WAN 10, CHA 10, GYA 11, KON 10, PEN 10, BIR 14, ZHA 07, SON 12, PAR 09, TUY 07, ZHA 13a, CHE 11, KOM 08, MEJ 11, EL 13, JIA 14, MEN 05, CHE 03, SZA 07, WAN 09].

2.2.5.2. *Normal-form and extensive-form*

Normal-form (strategic) game: A normal or strategic game is a game that takes place in a single turn in which the players are all simultaneously involved. It is represented by a table giving the players' gains for each possible outcome. Normal-form games include all possible combinations of strategies $s^1, s^2, ..., s^p$ such that p is the number of strategic combinations, and the payoff functions associating with each s^i the n-vector of payoffs $\mu^i, (i = 1, 2, ..., p)$. An example of an ordered coordination game is shown in Table 2.1, in the form of a matrix:

EXAMPLE.– *Follow-the-leader game*: in this game, two players X and Y try to decide whether the computers they sell will be designed to use small or large floppy disks. The two players will sell more computers if their disk readers are compatible. This means that, (If player X chooses large "L", choose L; if X chooses small "S", choose G), (If X chooses G, choose G; if X chooses P, choose P), If X chooses G, choose P; if X chooses P, choose G), If X chooses G, choose P; if X chooses P, choose P). This can be summarized as follows: "(G/G; G/P), (G/G P/P), (P/G G/P), (P/G P/P)".

		Y			
		Y1 (G/G, G/P)	**Y2** (G/G, P/P)	**Y3** (P/G, G/P)	**Y4** (P/G, P/P)
X	**X1** G=L	2,2	2,2E	-1,-1	-1,-1
	X2 P=S	-1,-1	1,1	-1,-1	1,1

Table 2.1. *Normal form of ordered coordination/payoffs of (X, Y)*

Extensive-form game: This is a game that is played in multiple turns. It is defined by a tree that describes how the game is played. In this case, each tree-top indicates the player who will choose an action at this stage of the game, as well as the information available to each player during decision-making. Each player's gains can be realized after having followed one of the possible routes within the tree from the roots to a terminal (leaf) node of the tree. It may also be the case that possible events and their probabilities are associated with certain terminal nodes. The extensive form of example 1 is shown below. Turns are taken simultaneously, which we indicate by letting X play the first turn while not allowing Y to know how Samir has played. The dotted line shows that Y's knowledge stays the same after X's turn. All X knows is that the game has reached a certain point within the information defined by the straight dotted line; he does not know the exact point that has been reached.

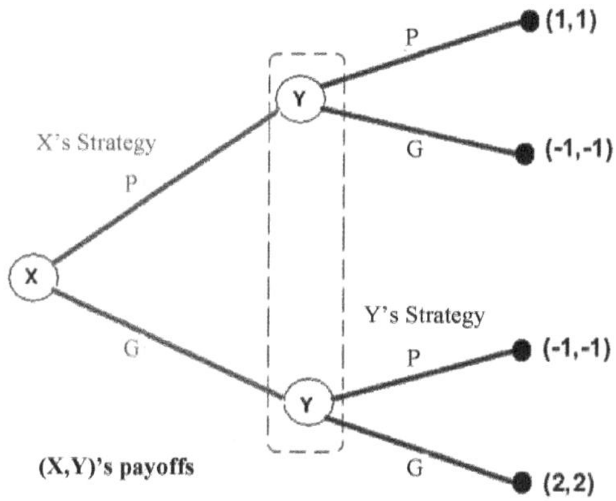

Figure 2.1. *Extensive form of ordered coordination*

2.2.5.3. *Perfect and imperfect information games*

When players are aware of everything that has already happened at the time they make their decisions, this is called a perfect information game. In this case, each node on the tree is visible to the players. However, when a player reaches a decision-making point at which he does not know the choices made by the other players who have taken their turn, this is called an imperfect information game. To represent a player's information in this case, we use the concept of a tree's information set. These sets indicate what a player knows at the time he takes his turn.

2.2.5.4. *Repeated games*

An ordinary game is a single game in which players decide simultaneously to take a certain action, while a repeated game is simply an ordinary game played several times in a row (with a finite number of repetitions). The conditions of the game are the same for all the repetitions of the game, including the same number of players, the same set of strategies for each player, the same gain and utility functions, and the same strategy update methods. Players determine their optimal strategies according to the history of the game. The repetition of several parts of the game allows players to coordinate their present and future choices with their past choices, which causes new solutions and *new equilibriums* to appear (folk theorem).

2.2.6. *Equilibrium*

The analysis of a game consists of predicting an equilibrium outcome between rational players. Equilibrium in a strategic game is a state or situation in which no player wishes to modify his behavior laterally, taking into account the behaviors of the other players. In other words, equilibrium is a combination of strategies such that none of the players wishes to change his strategy given the strategies of the other players. The games we are considering here correspond to games in which each player chooses his strategic changes alone without consulting the other players; thus we speak of non-cooperative games, which do not offer the possibility of influential structural cooperation between players. When we wish to determine equilibriums, we do not consider the incitement of groups of players to modify their behavior jointly, given the behavior of the remaining players. A model without equilibrium or with multiple equilibriums is incorrectly specified, or its author has not been able to supply a complete and precise prediction of what would happen.

We use notation s_{-i}, which represents the combination of all the players' strategies except that of player i ($i = X$, example 1). This combination is of capital interest for i, because he uses it to try to choose his own strategy. The new notation also helps him to define his best response [TOU 99]. If we use s_i^* to designate the equilibrium of player i, then an equilibrium s^*, which is a combination of strategies composed of one best strategy for each of the n players of the game, may be defined as follows:

$$s^* = (s_1^*, \ldots, s_i^*, \ldots, s_n^*) \qquad [2.1]$$

2.2.6.1. *Best response and dominant strategy*

To determine the equilibriums of a game, we start by eliminating all dominated strategies and then look for equilibriums within the reduced game. A rational player never admits a dominated strategy in terms of gains by at least one of his other strategies in the face of all his rival's possible strategies. The best response of player i to the strategies s_{-i} chosen by the other players is strategy s_i^*, which gives him the largest payoff, specifically:

$$\mu_i(s_i^*, s_{-i}) \geq \mu_i(s_i, s_{-i}), \qquad \forall s_i \neq s_i^* \qquad [2.2]$$

The best response is said to be strictly best if no other strategy is as good, and the best if there is at least one other strategy that is as good. Strategy s_i^* is a dominant strategy if it constitutes the strictly best response of a player to any strategy the other players might choose; in other words, the payoff associated with s_i^* is higher than the one associated with the choice by i of any other strategy. This can be written as:

$$\mu_i(s_i^*, s_{-i}) \geq \mu_i(s_i, s_{-i}), \qquad \forall s_{-i}, \forall s_i \neq s_i^* \qquad [2.3]$$

2.2.6.2. *Dominant strategy equilibrium*

DEFINITION 2.1.– *if a strategy is dominant, it is the best response to any strategy chosen by the other players, including their equilibrium strategies [TOU 99, KAC 12, RAS 01].*

So we can say that dominant strategy equilibrium is the combination of strategies including the dominant strategy of each player.

DEFINITION 2.2.– *When dominant strategy equilibrium exists, it is unique [TOU 99].*

This type of equilibrium gives us a very clear and intuitive prediction of the result of a game. Unfortunately, it exists only for very few games. Therefore, we must introduce other concepts of equilibrium in order to predict the solutions of different types of games. There are several equilibrium concepts, but Nash is the most applicable and the most widely used in game theory.

2.3. Nash equilibrium

2.3.1. *Definition*

John Nash developed a method to solve interactive games called "Nash equilibrium". This theorem is considered to be the most effective solution to various types of conflict games. The theoretician John Nash has demonstrated that it is possible in any contentious situation, under certain conditions, to reach a state of equilibrium, which leads to stability, in which all of the players will be satisfied with their gains and none of them seeks to change his situation. In Nash equilibrium, the player is not necessarily happy with the strategies of the other players, but his strategy is the best response to their actions. Players in Nash equilibrium are always non-cooperative. When a game model does not specify which equilibrium concept it is using, it will certainly be Nash's, or a refinement thereof. The way to approach equilibrium consists of proposing a combination of strategies and seeing whether each player's strategy is a best response to the others' strategies.

DEFINITION 2.3.– *Any dominant strategy equilibrium is a Nash equilibrium, but the opposite is not always true [XIA 08, TOU 99].*

DEFINITION 2.4.– *For a strategy to be a component of a Nash equilibrium, it must simply be a best response to the equilibrium strategies of the other players [TOU 99].*

By definition, the combination of strategies s^* is a Nash equilibrium, if the inequality below is satisfied for each player *i*, $i \in N$;

$$\forall i,\ \mu_i(s_i^*, s_{-i}^*) \geq \mu_i(s_i, s_{-i}^*), \qquad \forall s_i \in S_i \qquad [2.4]$$

or:

$$\mu_i(s_1^*, \ldots, s_i^*, \ldots, s_n^*) \geq \mu_i(s_1^*, \ldots, s_i, \ldots, s_n^*) \qquad [2.5]$$

That is, the gain of player i when he chooses s_i^* and all the others make a choice in compliance with s^*, is greater than the gain of the same player i when he deviates from s^* and selects another strategy s_i. In other words, if no one can benefit from a deviation from s^*, then no one will do it, which is proof that s^* is an equilibrium. Here is an example (*Example 2*) of a game described by the following matrix: If 1 plays A, it is optimal for 2 to play D; and if 1 plays B, it is optimal for 2 to play C. In addition, if 2 plays C, it is optimal for 1 to play B, and if 2 plays D, it is optimal for 1 to play A.

1 \ 2	C	D
A	(0,0)	(2,2)
B	(10,11)$_E$	(-1,0)

Table 2.2. *Payoffs of players (1,2)*

However, pair (B, C) seems to be a reasonable solution in the sense that no player can do better for himself. Therefore, this pair constitutes a Nash equilibrium.

Like a dominant strategy equilibrium, a Nash equilibrium can be weak or strong. The definition offered above concerns the weak version. To define a strong Nash equilibrium the inequality must be made strict; that is, it must ensure that no player is indifferent between his equilibrium strategy and another strategy. In addition, there are specific games that have multiple equilibriums, and other games that no longer have an equilibrium. In reality, a no-equilibrium or multiple-equilibrium model is poorly specified, which means that its author has not been able to supply a complete and precise prediction of what will happen. This is why the non-existence or absence of uniqueness poses a major problem in game theory.

2.3.2. *Existence*

There are conditions necessary for proving the existence of Nash equilibrium in a strategic game. To further explain this, we will start by defining the best response by the $\arg\max$ function. So, s_i is the best response of player *i*, $i \in N$, if;

$$r_i(s_i) = \arg\max\ \mu_i(s_i, s_{-i}), \qquad \forall s_{-i} \in S_{-i} \qquad [2.6]$$

THEOREM 2.1.– When the group of strategies of player i is a compact group, and when his gain is a continuous function of s_i; then a maximum exists. (Weierstrass).

DEFINITION 2.5.– *A digital application f defined on S is said to be strictly quasi-concave if;*

$$f(s_i, s_{-i}) > \min\{f(s_i^{'}, s_{-i}), f(s_i^{''}, s_{-i})\} \qquad \forall s_i \in \left]s_i^{'}, s_i^{''}\right[\qquad [2.7]$$

We also say that indifference curves are strictly convex in relation to the source.

If μ_i is strictly quasi-concave in s_i, then the gain function admits a single maximum. Supposing on the contrary that $\mu_i(s_i, s_{-i})$ possesses two maximums, $s_i^{'}$ and $s_i^{''}$ for a given s_{-i} is ordinary. Since μ_i is strictly quasi-concave in s_i, we have:

$$\mu_i(s_i, s_{-i}) > \min\{\mu_i(s_i^{'}, s_{-i}), \mu_i(s_i^{''}, s_{-i})\} \qquad \forall s_i \in \left]s_i^{'}, s_i^{''}\right[\qquad [2.8]$$

so that s_i guarantees gains that are strictly higher than the maximum gain, which is a contradiction. Consequently, under the preceding hypotheses, $r_i(s_{-i})$ is an application defined for the group S_i, meaning that a better response exists, no matter what strategies are chosen by the other players.

THEOREM 2.2.– If utility function f is concave then an equilibrium exists for this function. (Rosen)

Most proofs of the existence of Nash equilibrium are based on the fixed-point theorem.

DEFINITION 2.6.– *We say that an application* $f: X \rightarrow X$ *, admits a fixed point, if* $x_0 \in X$ *is its own image by* $f; x_0 = f(x_0)$.

Consider the application $r(s): S \rightarrow S$, such that; $r(s) = \{r_1(s_{-1}), \ldots, r_n(s_{-n})\}$, so;

THEOREM 2.3.– If application $r(s)$ possesses a fixed point, then this point is a Nash equilibrium of the game, and vice versa.

PROOF.– Suppose that s^* is a Nash equilibrium, so for any $i = 1, \ldots, n$ we have $s_i^* = r_i(s_{-i}^*)$, which implies that s^* is a fixed point of $r(s)$. In the opposite direction, Si s^* is a fixed point of $r(s)$. This means that $s_i^* = r_i(s_{-i}^*)$, for any $i = 1, \ldots, n$. So; $\mu_i(s_i^*, s_{-i}^*) \geq \mu_i(s_i, s_{-i}^*)$, which means that s^* is a Nash equilibrium.

A set of sufficient conditions for a fixed point to exist is given by the following theorem:

DEFINITION 2.7.– *Let application* $f: X \rightarrow X$, *where* X *is a subset of* R^n . *If* X *is compact and convex and if f is continuous, then f possesses a fixed point. (Brouwer).*

THEOREM 2.4.– If sets of strategies are compact and convex subsets of R^n , and if gain function μ_i is continuous in s and strictly quasi-concave in s_i for each player $i = 1, \ldots, n$; then the non-cooperative game admits a pure-strategy Nash equilibrium (Debreu, Glicksberg and Fan) [KAC 12].

In addition, if we remove the strictly quasi-concave hypothesis and replace it simply by quasi-concave, there will be more uniqueness of the maximum. Moreover, the continuity of μ_i in relation to s_i and the compactness of S_i are essential for guaranteeing the existence of a maximum of μ_i in S_i. The strict quasi-concavity of μ_i *in relation to* s_i and the convexity of S_i imply the uniqueness of the maximum. All these conditions lead us to state that best-response curves are continuous, and consequently they intersect with each other.

THEOREM 2.5.– Any finished game admits a mixed-strategy equilibrium. (Nash) [RAS 01].

To sum up, equilibrium does not exist when the utility function is discontinuous. Knowing that the utility function of pure strategies is discontinuous implies that the utility function of mixed strategies is also discontinuous.

2.3.3. *Uniqueness*

DEFINITION 2.8.– *A function* $f : X \rightarrow X$, $X \in R^n$ *is a contraction if* $\lambda \in [0,1]$ *exists, such that for any x' and x", we have:* $d\left[f(x'), f(x'')\right] \leq \lambda d(x', x'')$.

Meaning that the images are closer than the starting points.

THEOREM 2.6.– If $r(s)$ is a contraction, then the Nash equilibrium is unique.Thus, if the best-response curve is continuous and of a slope smaller than unity, then $r(s)$ is a contraction.

Remember here that the uniqueness and existence of Nash equilibrium are the most difficult points to study in game theory. Moreover, experiments have shown situations in which the authors have failed to demonstrate either the existence or even the uniqueness of pure-strategy Nash equilibrium, and they were obliged to bring external factors into play or, in the worst-case scenario, to remodel the game.

2.3.4. *Specific cases*

There are some specific games that possess multiple equilibriums, and others that do not possess an equilibrium. For multiplicity of equilibriums, the most famous example is probably that of the "battle of the sexes" (*Example 3*). This is a conflict between a couple; a man who wants to go to a football game (M) and his wife, who wants to go to the theater (T). The game is summarized in the matrix below:

	F \ H	T	M
3/4	T	$(3,2)_{E1}$	(1,1)
1/4	M	(0,0)	$(2,3)_{E2}$

Table 2.3. *Two equilibriums in the battle of the sexes*

Note that there are two Nash equilibriums given by (T,T) and (M,M). Here, we must recognize that the game specification is incomplete. In this case, external factors are liable to play an important role in the emergence of a particular solution. For this, in the battle of the sexes, for example it will be useful to know who acts first. If the man could buy the ticket for the game in advance, his commitment would motivate the woman to go to the game, and we would be at a single equilibrium E2. We can also see that in example 1, summarized in Table 2.1, payoffs E1, E2, and D3 constitute Nash equilibriums, such that E1 and E3 are not quite reasonable, but we must solve this mathematically to retain a single equilibrium E2 at the end.

The equilibriums discussed above are equilibriums taken from pure strategies. However, we may decide to use a random mechanism that decides for the players. To do this, we can remodel the game so it appears clearer, by giving each player a subjective probability. Thus, we suppose that each player associates a probability p_i with the pure strategy s_i and leaves it up to the random mechanism to decide. Now, each player tries to maximize his hoped-for gains by choosing the best possible combination, which is called *mixed strategy*. Let us go back to *Example 3*, in which we found two pure-strategy Nash equilibriums. There is a third Nash equilibrium in mixed strategy, to which both players can turn in order to avoid the indeterminacy associated with relying on single pure strategies. Thus, player (**H**) will have, for example, a probability y of choosing (T) and a probability (1-y) of choosing (M). The same is true for player (**F**), who will have a probability x and (1-x). Consequently, player (F), who prefers the theater, puts a higher probability on (T) than on (M), and vice versa for player (H).

There are also games that do not possess a pure-strategy Nash equilibrium. The most famous example in this category is "matching pennies", which takes place between two players, who simultaneously call heads or tails. This game is a *zero-sum game*, which means that the total gain at the end of the game is nil (the gain won by one of the players is equal to the gain lost by the other). This game does not have a pure-strategy Nash equilibrium. So, to bring out an equilibrium solution, we can proceed as in the previous example by introducing mixed strategies and remodel the game again.

2.4. Famous games

2.4.1. *The prisoner's dilemma*

The prisoner's dilemma is the best-known example in game theory. Its premise is as follows: "The police arrest two suspects who have committed a crime together, and question them separately. They present each of them with the following demand: 'If your accomplice confesses (A) and you keep silent (T), you'll get a firm ten-year sentence and he'll go free. If the opposite happens, you will be able to go free but he'll rot in prison. Otherwise, if you both confess, the sentence will be shared (five years).' If both prisoners keep silent, the sentence will be three years for each of them."

The possible choices of the two prisoners (P1 and P2) can be represented as follows:

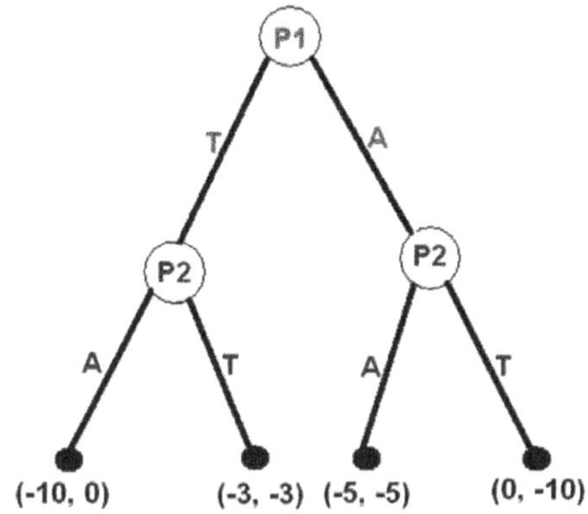

Figure 2.2. *Prisoner payoffs*

Nash equilibrium is reached in this game when both prisoners confess (confess, confess).

2.4.2. *Cournot duopoly*

"*Let two firms A and B be producers of the same very low-cost product, with infinite production capacities. The profit from the product of these two firms will result from the total quantity put on the market.*"

This is a strategic perfect-information game. The solution, according to Cournot, is for a pair to offer (s_1, s_2) for (A, B) respectively, such that each firm maximizes its profit, given the other's offer.

So, $i = 1,2$. The strategy is the quantity of the product that each firm produces, which equals $S_i = [0, +\infty[$. The gains (utilities) of each player are: $\mu_i(s_1, s_2),\ s_i \in S_i$

such that:

$$\mu_1(s_1, s_2) = [p - (s_1 + s_2)]s_1 - c(s_1)$$

and:

$$\mu_2(s_1, s_2) = [p - (s_1 + s_2)]s_2 - c(s_2)$$

So, if:

$$\frac{\partial \mu_1}{\partial s_1} = 0 \Rightarrow p - 2s_1 - s_2 - c = 0 \Rightarrow s_1 = \frac{p - s_2 - c}{2}, \quad s_1 = f(s_2)$$

and:

$$\frac{\partial \mu_2}{\partial s_2} = 0 \Rightarrow p - 2s_2 - s_1 - c = 0 \Rightarrow s_2 = \frac{p - s_1 - c}{2}, \quad s_2 = f(s_1)$$

At equilibrium (s_1^*, s_2^*) we have:

$$s_1^* = f(s_2^*), \text{ and } s_2^* = f(s_1^*)$$

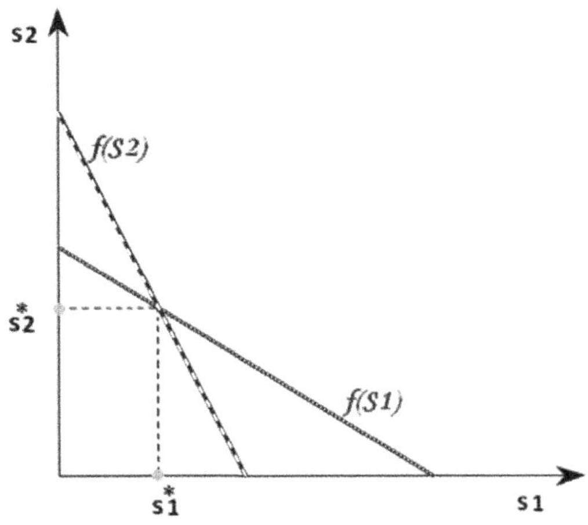

Figure 2.3. *Nash–Cournot equilibrium*

So Nash–Cournot equilibrium (s_1^*, s_2^*) is the intersection of two reaction function curves $f(s_1)$ and $f(s_2)$.

2.5. Applications to wireless networks

In addition to its applications in various fields including economic, political, and social sciences, biology, and information technology, game theory has shown its extreme usefulness in the field of telecommunications, in wireless networks in particular, for which it provides effective solutions to several existing issues. Game theory offers researchers a highly powerful mathematical tool that can be used to model very complicated situations that appear during the operations of these networks, taking into consideration all the factors and elements at work in these situations.

In wireless networks, communications occur over a radioelectronic medium based on the free propagation of electromagnetic waves in space. This natural medium is common and shared by all users of the network, which can cause interference between signals transmitted simultaneously by the various transmitters using the same physical link. In addition, though access to the transmission channel is managed by fixed access techniques such as time division multiple access (TDMA), frequency division multiple access (FDMA) and code division multiple access (CDMA), or is random access techniques such as random access protocol (ALOHA) and carrier sense multiple access (CSMA), collisions between packets still occur, and have serious effects on the quality of service (QoS) of the network. In reality, phenomena of interference and collision indicate to us that there are interactions and conflicts between the network's users. All of this explains the importance of using game theory, the objective of which is to converge toward an effective allocation of resources, energy control and output optimization. This has been proven in recent years by the fact that resource allocation based on game theory has considerably improved effectiveness in the exploitation of the radio spectrum.

Based on these ideas, we can pick out numerous situations in which game theory can be involved in the form of different strategic game scenarios. If network users play the role of players then we can have multiple strategies, such as power level, transmission speed and routing nodes. We can also predict the following utilities: bit error ratio (BER) attained, energy level conserved, time elapsed and transmission output. Moreover, in the layers of the network, and particularly in the physical and medium access control (MAC) layers, the use of game theory is highly beneficial. Moreover, the MAC and Physical (PHY) layers have experienced significant advancements following multiple studies and research products conducted on their basic functions. For the upper layers of the network there are several competitive situations, such as the case, for example, of competition between the various operators and service providers in cellular networks [MAC 08], and also the bids operators must make for the allocation of radio resources, etc. Thus it is clear that wireless networks are full of phenomena that can be addressed using game theory.

In the PHY and MAC layers, the atmosphere seems quite sufficient for the exercise of game theory. This is due to the various conflict situations that these two layers experience; for example, in the MAC layer, users share the same channel to access the environment, the same routes for the delivery of packets, the same routers, and sometimes even the same transmitting or receiving nodes. Interactions in situations like this are established and can negatively affect the network's yield. However, for circumstances like these, the use of game theory is quite useful for planning convenient solutions that will lead to the stability and optimization of the network. We will now give some practical examples of the use of game theory in different services offered by wireless networks, and then provide some examples of game models for wireless networks.

2.5.1. *Routing game*

The fact of sharing routes and routing points among network users creates an atmosphere of competition and interaction among them. Routing is among the network functions most analyzed by game theory in recent years. In fact, many articles in the literature examine routing in the form of several game models; for example the congestion and potential games described in [ROS 73] and [BAR 08], respectively. In addition, Wardrop has introduced a new game model in [WAR 52] for the organization of network traffic in order to improve the network in general in terms of time and speed. This model is currently being used as a reference in several studies on routing in communication networks.

To facilitate a better understanding of this application we will now introduce a game model proposed in [ROS 73, MOD 96], which introduces a new routing protocol in *ad hoc* networks. The author presents a protocol that sends each packet from each source via the shortest route toward its destination. Routing decisions are made in mobile nodes that sometimes serve as routers. Route choices are made according to game equilibriums (Wardrop equilibrium). In this incomplete-information model, the node takes the role of the player and the strategy of each player is to determine the nodes toward which he will transport his tasks, and utility represents the time, in that each player seeks to find the shortest path (fewer nodes) to take the smallest amount of time. The game is presented formally by a graph $G = (V, E)$, such that V is the set of nodes and E is the set of arcs. Player i, $i \in N$ is characterized by:

– its weight w_i which represents the size of the message to be transmitted;

– a pair of peaks $(s_i, t_i) \in (V \times V)$ which represent its source and its destination;

– a set p_i, the shortest path between s_i and t_i; and

– a strategy vector q_i indexed on the routes of p_i. For any $j = \{1,...,m\}$, q_{ij} is the probability for i to choose j in p_i.

During the game, each player i chooses a path p_i.

Utility: The cost of a given path is the sum of the costs of the arcs that compose it. Thus, for any path p the associated cost C_p is defined as follows, such that $e \in E$:

$$C_p = \sum_{e \in p} c_e \left(\sum_{i=1}^{n} 1_e \times w_i\right) \qquad [2.9]$$

Such that, 1_e equals 1 if $e \in p_i$, and 0 if not. Utility is defined as follows:

$$U_p = 1 - C_p \qquad [2.10]$$

A player chooses a path in each section to transmit its message. The updating of its strategy results in an increased future probability of taking this path. The greater the utility of the path (the lower its cost), which causes a significant increase. All the other probabilities are reduced by a certain proportion, such as the utility function of the path selected. Finally, the author specifies a learning method based on the path vectors and Lyapunov's stability theory, which enables the system to converge toward a stable state, which is a Nash equilibrium.

2.5.2. *Power control game*

As we saw in Chapter 1, the management and control of energy in *ad hoc* network is of vital importance. The nodes are equipped with batteries whose capacity is very limited, which requires the implementation of an effective policy to monitor and save energy. To do this, distributed algorithms whose objective is to minimize the power transmitted, thus ensuring acceptable signal to interference plus noise ratio (SINR) and QoS have been proposed [KOS 05, CHE 13]. There are conflicts of interest between mobile phones due to interference, such that if one mobile phone increases its power then it will increase its SINR, but on the other hand, it will weaken the SINRs of the others [XU 12, ILT 06]. So, we can model these conflicts as a non-cooperative game with imperfect information. Here, the link acts as a player, the power is the strategy used by the player, and the objective of the game is to keep an SINR level t_i higher than a certain threshold t_0 in order to ensure an adequate QoS. We use p_i to indicate the transmission energy of player i, such that $p_i^{\min} < p_i < p_i^{\max}$, and the energy vector of the N players is $p = (p_1, \ldots, p_N)$. The SINR of player i is:

$$t_i(p) = \frac{p_i g_{ii}}{n_i + \sum_{i \neq j} p_j g_{ji}} \qquad [2.11]$$

Such that, n_i is the route noise of player i and g_{ji} is the gain of the link between the transmitter of player i and the receiver of player j. Also, the output T_i of player i is given as follows:

$$T_i(p) = W \log_2(1 + \frac{t_i(p)}{\tau}) \qquad [2.12]$$

Such that W is the bandwidth, and $\tau \geq 1$.

The usefulness for each player i will be defined as follows:

$$\mu_i(p) = \begin{cases} \frac{T_i(p)}{g_i}, & \text{if } t_i(p) \geq t_0 \\ 0, & \text{other} \end{cases} \qquad [2.13]$$

The transmitter i is accepted if; $t_i \geq t_i^*$, so:

$$p_i \geq t_0 \left[\sum_{i \neq j} \frac{g_{ij}}{g_{ii}} p_j + \frac{n_i}{g_{ii}} \right] \qquad [2.14]$$

We write, $\frac{g_{ij}}{g_{ii}} = h_{ij}$, and $\frac{n_i}{g_{ii}} = \eta_i$. If we take the example of two players only, we have:

$$\begin{cases} p_1 = t_0(h_{12}p_2 + \eta_1) \\ p_2 = t_0(h_{21}p_1 + \eta_2) \end{cases} \qquad [2.15]$$

In his report, the author presents a game scenario composed of N pairs of terminals, where each pair acts as a player, which results in an N-player game.

The study has shown that this algorithm converges toward a single solution that presents a Nash equilibrium for this game. Consequently, at equilibrium, players obtain perfect power levels with a smaller convergence time. In fact, this study has also shown us that the correct determination of the utility function is an extremely important part of the model's success, and enables us in the end to have more practical results.

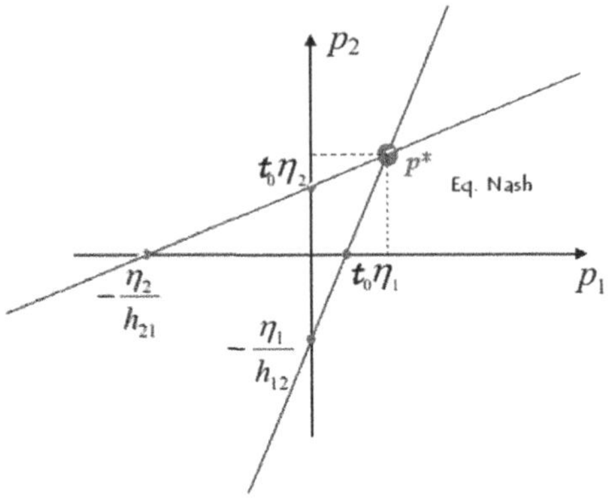

Figure 2.4. *Nash equilibrium in a two-player power-control game*

2.6. Conclusion

Game theory is a highly effective mathematical tool for the analysis of conflict situations that involve interactions between their decision-making elements. These types of situations are analyzed via the supposition of a model in the form of a strategic game, for which the precise definition of each of its elements is vital. However, the study of equilibrium remains the most important and most difficult part of game analysis. For this reason, in this chapter we have introduced all the rules and conditions sufficient to demonstrate the convergence of the model and thus discover the game's equilibrium, if it exists.

With regard to the many competitive and interactive situations it presents, a wireless network is considered to be a favorable climate for the application of game theory. Moreover, a great deal of study and research has already been conducted on their various functions based on strategic game models. In the second part of this chapter, we demonstrated how game theory is useful for wireless networks, illustrating this point with practical examples of strategic games applied to wireless networks.

3

Games in SALOHA Networks

3.1. Introduction

The ALOHA protocol and its improvements have been widely applied in satellite and terrestrial cellular networks. It was developed by a laboratory at the University of Hawaii in 1970 [ABR 70] to establish a radio link between a central station and other mobile stations. It is generally designed for low traffic load transmissions with a small number of users. Moreover, in the frequent event of a large number of inactive sources, random access enables us a more efficient use of the channel. For satellite networks, it can be used either to transmit data or to make reservations by requesting to be assigned a fixed frequency band. The protocol studied in this chapter is Slotted ALOHA (SALOHA) [ROB 72], which is an advanced version of classic ALOHA , in that time is divided into subintervals of equal duration called slots. The fundamental principle of this algorithm is based on random access to the transmission channel, such that each node in the network transmits the first packet to arrive in the first slot, and then the next packet arriving in the second slot, and so on for the other packets, always according to the first input first output (FIFO) mechanism. The advantage of this method is that there can be a shorter time for the transmission of packets. However, when another node in the network transmits at the same time as another node in the midst of transmission, collisions can occur between packets, and transmission fails as a result. The situation becomes even more complex when multiple nodes wish to transmit simultaneously.

In SALOHA, transmission always begins at the start of the slot, in synchronization with the system clock. The node confirms the success of its transmissions via the reception of an Acknowledgment (ACK) signal, which is considered an acknowledgment of reception of transmitted packets. However, if the node does not receive the ACK signal at the end of the current slot, it knows that the packets have collided and so the transmission has failed. In addition, if packets that have collided – also called "holding packets" – are retransmitted immediately in the next slot, other collisions will surely result. To avoid more collisions, these packets will be put on hold for a random amount of time and then retransmitted. This method must be respected by all nodes until all packets have been successfully transmitted.

Upon initial observation, we can see that SALOHA becomes quite weak in the event of multiple spontaneous use of the transmission channel, meaning that it has low immunity against packet collision problems. Transmission time is also a critical factor in the operation of this type of network. A node may find itself in collision several times, and a considerable amount of time may be spent before it finishes sending all of its packets. However, when there are only a small number of nodes, the network becomes more stable and presents shorter transmission times. It is also clear that the holding period before retransmission is critical and that, whether it is short or long, it affects the network's output and quality of service (QoS). However, a good decision on the length of this waiting period can lead to the lining-up of transmissions of packets on hold. Nodes are then required to choose the best slot times either to begin or to continue their transmissions; otherwise, the number of collisions will multiply, as series of retransmissions are added back into the network, causing prolonged transmission times and reducing network performance in terms of output and QoS. Also, in conditions like these, the most limiting problem is that of energy; when large quantities of energy are lost gratuitously during successive packet retransmissions, which can even prevent the completion of missions planned during the setup of the network [ZHA 12, LI 14, GÁL 12].

Thus SALOHA, with its rules of operation, creates an atmosphere of conflict and interaction between network nodes, setting up a kind of competition between different nodes in the process of transmission, which seek to collect free slots that they will use later to complete their transmissions. We might say that SALOHA motivates nodes to act egotistically, with each protecting its own benefits, which take the form of transmission time or output, such that each node tries to finish the transmission of all its packets in as short time as possible and using the smallest amount of energy, all without considering the behavior or interests of the other nodes. This places us in a non-cooperative, imperfect-information strategic game. In situations like this, game theory seems to be the best analytical tool with which to reach a certain level of equilibrium, in which each node will be satisfied in terms of its interests and the game (network) will function in stability with significant optimization. In recent years, SALOHA has been the subject of a great deal of scientific research, particularly using game theory to model SALOHA networks as strategic games with or without cooperation, and with perfect and imperfect cooperation, all with the objective of making this network stable and more reliable.

In this chapter, we will look at wireless networks based on random access protocol: SALOHA. We will also present our strategic non-cooperative game model with imperfect information as a real scenario. In reality, nodes in wireless networks, and notably in mobile *ad hoc* networks (MANETs), do not form any sort of cooperation or exchange of information with one another, which means that the model we will be presenting here is more practical in its characteristics and reflects the network in a real scenario. This gives our results the advantage of being more efficient. The objective of our study is to discover the optimization level in transmission throughput, energy consumption and quality of service, localized in the convergence network zone and equilibrium point.

This chapter is structured in two main parts, each part dedicated to the presentation and detailed description of our strategic game model. In the first part, we will present a classic study of SALOHA based on the use of Markov chains to determine the various operating states of this algorithm, and thus to define the throughput function of the network. Next, we will move on to the definition of our strategy game, called a *coding game*, and will define the principal elements that compose it as well as its various rules and characteristics. The study of game equilibrium is the most important and the most difficult part of game theory. Here, we will explain in detail (using a set of mathematical equations and developments) how the model converges under certain conditions toward a unique solution representing Nash equilibrium (NE). In fact, we have proved that our game is convergent, and accepts a unique solution that presents the point of NE. We will conclude this part with simulations and comparisons to demonstrate the advantages of the model proposed in comparison to a conventional model.

The second part of this chapter is focused on SALOHA performance at equilibrium. We will use our game's NE for the SALOHA network in order to assess the level of optimality it may contribute to various network performances in terms of coding value, loss rate, and transmission time and output, as well as energy consumption by nodes, something that can be clearly seen through simulations and comparisons made to the conventional model.

3.2. Functioning of the SALOHA algorithm

Unlike classic ALOHA, in SALOHA time is discretized into subintervals of equal duration, called slots (time slot: TS), and transmissions are synchronized so that they begin at the start of TS, which means that the period of vulnerability will be half of the one in classic ALOHA. Consequently, the number of collisions will be reduced, eliminating the collisions that occur in classic ALOHA when a node starts transmission during the vulnerability interval of another node in the midst of transmission. In addition, in SALOHA the system is on discrete time, which further impacts analysis and simplifies study. Our study in this chapter is focused on the SALOHA-based wireless network as being the medium access technology. The network is composed of M active nodes that share a transmission channel toward a receiving station in single-hop mode, as shown in Figure 3.1. Before moving on to mathematical developments, however, we must first consider the following suppositions:

1) The packets being transmitted are of equal size or length, and each of them initially requests a single TS for its transmission.

2) All transmissions are synchronized in relation to the system clock, so that each transmission begins at the start of a new TS and finishes at the end of the same TS.

3) The arrival of each node's packets follows a Poisson process of average value λ, such that $\lambda/_M$ is the transmission speed of each node.

4) When a node transmits alone during a TS, then the packet will be received correctly; but when multiple nodes transmit simultaneously, collisions will occur successively and none of the packets will be correctly received. Note here that according to this proposition, and because real distances between transmitters and receivers in a MANET are generally small, in our analysis we will disregard errors made due to noise and the capture effect, which can occur during multiple transmissions.

5) At the end of each TS occupied by a transmission, the node will receive an information packet or feedback (ACK) from the receiving node giving information on the state of reception (one packet received, zero packet received or more than one packet received in case of error). This supposition is far from the (non-practical) reality in the case of a satellite network.

6) A packet that has been in a collision will be considered for subsequent retransmission after a random time period (several TS), and this mechanism is repeated until the successful transmission of the packet. We call the node containing packets to be retransmitted, a holding node.

7) A holding node diverts newly arrived packets until it is finished transmitting its current packets (without storage). In the event of low traffic, the number of retransmitted packets (on hold) will be

negligible, which means that the storage effect no longer affects the system in this case.

8) The operating principle of SALOHA is based on the fact that active node, which is not on hold, immediately transmits the new packet received in the first available slot. Though this method risks creating collisions in the event of simultaneous transmission, it has the advantage of shorter times in the event of few collisions. In addition, if collided packets are retransmitted immediately in the next slot, this will undoubtedly cause more collisions. In order to avoid this, the holding node always delays its retransmissions by a few slots.

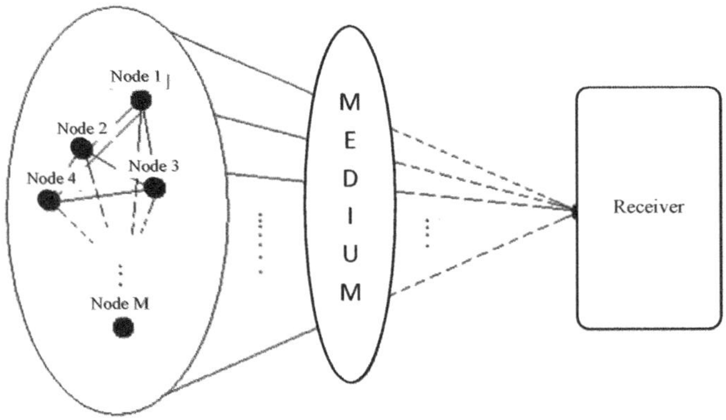

Figure 3.1. *Model of wireless network with M nodes*

To analyze this algorithm, we will consider the model shown in Figure 3.1, which presents a wireless network composed of M nodes sharing a transmission medium toward a receiving station. New packets transmitted by slot adhere to a Poisson process of average value λ. In addition, slot retransmissions of packets on hold form a Poisson process of average value G, such that $G>\lambda$. By this approximation, the probability of a successful slot transmission will be equal to Ge^{-G}.

We can see that the maximum output (0.368 packets/TS) is reached at value $G=1$, which is logical in practice, while in reality there will be no collisions at this transmission rhythm and the network will function in an ideal manner. Moreover, outside of this maximum value, we can see that each output level can theoretically be reached by two different traffic loads (G_1, G_2); this is due to two principal causes; such that when $G < 1$, there will be many free slots generated, while for $G > 1$ the possibility of having a collision at that point increases, consequently reducing the number of successfully transmitted packets.

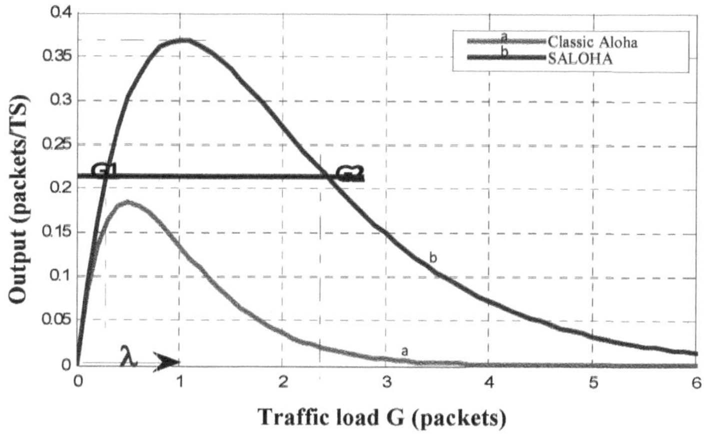

Figure 3.2. *Comparison of output between SALOHA and classic ALOHA*

However, the restrictive problem in SALOHA is that retransmissions of packets on hold coincide with new transmissions following a poor decision by choosing a slot that is already occupied by new transmissions, something that must be taken into serious consideration when analyzing the algorithm. To this end, the author in [BER 87] gives a more precise model of SALOHA, in which decisions of packet transmission and retransmission are probabilized. We suppose, then, that holding nodes retransmit with probability q_r in the next slot and until successful transmission. The number of slots bypassed before beginning the retransmission of holding packets, which was previously determined randomly, now forms a geometric random variable of index $i \geq 1$ with probability $q_r(1-q_r)^{i-1}$. Consequently, we can describe the behavior of SALOHA using a Markov chain [BIN 05, LIU 11, SHE 12, KON 05] with discrete time, in which the system state is the stochastic process with discrete time and in discrete state *n*, which represents the number of packets on hold. Each of these packets will be retransmitted independently with probability q_r, while the $M - n$ nodes remaining, which represent nodes that are not on hold, transmit their packets in consecutive slots. The arrival of packets in this scenario forms a Poisson process with an average value of λ/M, and an arrival probability equal to $e^{-\lambda/M}$.

Let $\pi(q)$ be the stationary distribution of the Markov chain, such that, in state *n*, $\pi_n(q)$ is the probability that *n* packets are on hold. Thus, the probability for a node that is not on hold to be transmitted in a given slot is:

$$q_t = 1 - e^{-\lambda/M} \quad [3.1]$$

If $Q_t(i,n)$ is the probability that *i* new nodes will transmit in a given slot, then:

$$Q_t(i,n) = \binom{M-n}{i} (1-q_t)^{M-n-i} q_t{}^i \quad [3.2]$$

And if $Q_r(i,n)$ is the probability that i nodes on hold will retransmit in a given slot, then:

$$Q_r(i,n) = \binom{n}{i} (1-q_r)^{n-i} q_r{}^i \quad [3.3]$$

3.2.1. *Study of stability*

In moving from one slot to the next slot, the system state (number of packets on hold) grows larger, via the addition of new arrivals of packets transmitted by new nodes. We write $Q_t(1,M) = 0$, and $Q_r(1,0) = 0$.

So, the probability of the system's transition from a state n to a state $n + i$ is given by the transition matrix below:

$$P_{n,n+i} = \begin{cases} Q_t(i,n), & 2 \le i \le M-n. \\ Q_t(1,n)[1-Q_r(0,n)], & i=1. \\ Q_t(1,n)Q_r(0,n)-Q_t(0,n)[1-Q_r(1,n)], & i=0. \\ Q_t(0,n), & i=-1. \end{cases} \quad [3.4]$$

The transition matrix of SALOHA shows how the system transitions from one state to another according to the different values of i. We can represent the overall transition of the system graphically as follows:

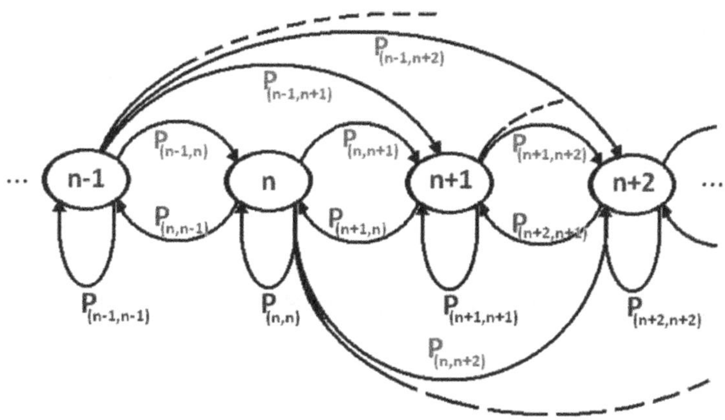

Figure 3.3. *Markov chain for SALOHA (graphic representation)*

We will now examine the behavior and performance of our SALOHA system with regard to traffic load and network characteristics. To do this, we will use $G(n)$ to define the probable overall number of slot transmissions, thus:

$$G(n) = (M-n)q_t + nq_r \quad [3.5]$$

We write A_n for the difference between two system states to represent explicitly the new accumulation of holding nodes created from one state to another, and which will be added to the future state. In other words, A_n is the number of new packets arrived in and accepted by the system minus the packets successfully transmitted in state n, such that:

$$A_n = (M-n)q_t - P_{succ} \qquad [3.6]$$

A packet has been successfully transmitted if it arrives alone at the TS without packets on hold, or if no new packets arrive and a single packet on hold will be retransmitted. Thus, the probability that a transmission will be successful is defined as follows:

$$P_{succ} = Q_t(1,n)Q_r(0,n) + Q_t(0,n)Q_r(1,n) \qquad [3.7]$$

And, if $q_t \ll$ and $q_r \ll$, we can simplify our calculations by using some approximations, such as $(1-x)^y \approx e^{-xy} / x \ll$, in equations [3.2] and [3.3], knowing that it is still considered that $q_r > q_t$ (the case of interest in our study). So, equation [3.6] becomes:

$$P_{succ} = G(n)e^{-G(n)} \qquad [3.8]$$

Thus, the probability of success in SALOHA is a function of q_t and q_r.

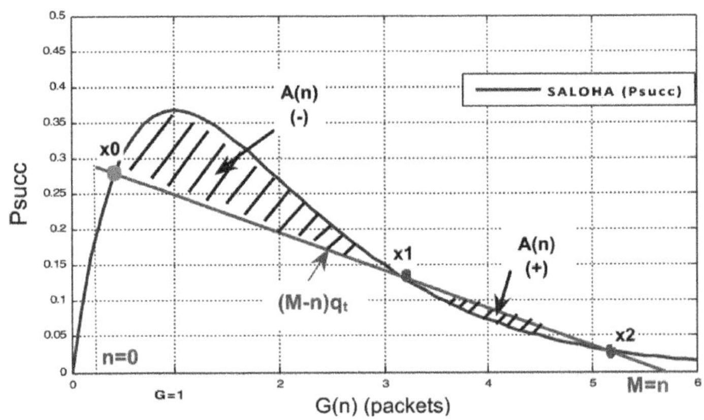

Figure 3.4. *Instability of SALOHA in probability of success and change of accumulation*

Based on Figure 3.4, we can see that the probability of successful transmission in SALOHA is maximized to $G=1$ when n is small, such that $G(n)$ is a function of n according to equation [3.5]. Also, accumulation between two successive system states is zero at three different points x_0, x_1, x_2. These points are the intersection of curve P_{succ} and straight line $q_t(M-n)$, which represent the equilibrium states of the system. Though the two points x_1, x_2 offer an accumulation that is theoretically nil, they cause a serious reduction in the network's efficiency and throughput since the probability of success is extremely low with a very large number of packets on hold, even tending toward the total number of packets $n = M$. Thus, the two points x_1, x_2 are undesirable points for the system and correspond in practice to a congestion situation (low output and long times), which reduces the system's efficiency. However, even while approaching x_0, which also presents accumulation that is theoretically nil, we can see a negative change in accumulation and will therefore attain a better P_{succ} than before, and which may even take the network to its optimum level $G(n)=1$. So, point x_0 is a point of stability for SALOHA, contributing efficiency and optimality from an output perspective. In summary, we can see that an unstable (bistable) functioning exists for SALOHA, which has a negative impact on the network's overall performance.

According to hypothesis (g) categorized above, which considers nodes in *no storage* mode, we thus obtain:

$$G(n) = \lambda + nq_r \qquad [3.9]$$

It remains the case that n is unknown (non-fixed), and since $G(n)$ is also a function of q_r, $G = f(q_r)$. So, to keep the system in stability (close to x_0), it is in our interest to set $G(n)=1$ while appropriately varying the value of q_r. According to equation [3.9], for example, we can see that when q_r is sufficiently reduced, and it still remains the case that $q_r > q_t$, then curve Ge^{-G} will spread slightly, so that the two undesirable points x_1, x_2 will disappear, and in the end only a single point of stability x_0 will remain. We will examine this later in this chapter.

3.2.2. *Transmission time*

A packet's transmission time is defined as the time necessary to transmit a packet from the time of its arrival at the node (transmission) until the successful transmission signaled by the reception of the ACK feedback. This time thus includes the sum of the transmission and retransmission times, as well as the time taken by feedback:

$$D = D_{trans} + D_{retrans} + D_{fback} \qquad [3.10]$$

So, we write [CHE 98]:

$$D = \frac{e - \frac{1}{2}}{1 - \lambda e} + \frac{(e^{\lambda} - 1)(e-1)}{\lambda\left[(e-1)(e^{\lambda} - 1) - 1\right]} \qquad [3.11]$$

Such that, in comparison to the model shown in Figure 3.1 and supposing a single receiving station, the time taken by feedback will be considered negligible in the face of the total transmission and retransmission times, such that $D_{fback} \ll D_{trans} + D_{retrans}$, knowing that in SALOHA, so as not to complicate the transmission, the ACK signal is assumed to be transmitted via a separate channel with a probability of 1.

We can see in Figure 3.5 that when the number of packets arriving in the network increases, the time increases relatively up to a value of (1/e = 0.3679), when it suddenly increases very rapidly toward an unlimited value. This can be explained by the fact that if the number of packets increases but always remains lower than 1/e $\lambda \leq \frac{1}{e}$, the system functions in its stability zone and has short transmission times with the availability of free slots and despite the generation of a few packets on hold; these packets will be retransmitted subsequently in the next slots, thus taking up short periods of time. However, when the traffic load exceeds 1/e ($\lambda > 1/e$), the system becomes unstable and takes long periods of time to transmit the arriving packets. In fact, value (1/e) corresponds exactly to point x_0 in Figure 3.4, which represents the highest performing level of stability for SALOHA.

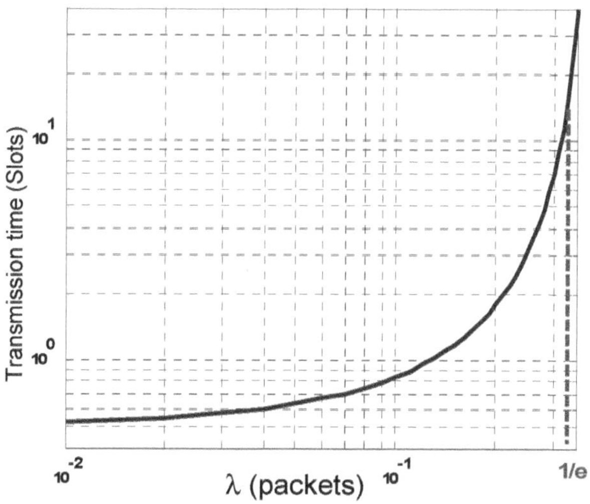

Figure 3.5. *Transmission time of a packet*

3.3. Modeling of node behavior in SALOHA with a strategic coding game

3.3.1. *Issues*

During operation, a SALOHA network generally shifts between two crucial temporary states: a state of progression followed rapidly by a state of saturation. In the state of progression, the number of nodes wishing to transmit is relatively small given a transmission channel offering a sufficient number of free slots. This increases the probability of successful transmission and also improves network performance in terms of output and transmission time. However, when active nodes arrive, the network begins to experience collisions and congestion in the transmission support, and very rapidly shifts to a state of saturation, in which network performance will be seriously reduced. Note that there is no middle state of stability between the two transitory states, which might ensure a certain level of reliability and efficiency during prolonged periods of network operation. It is in this context that our study is focused, as we believe that through our strategic game model we have discovered a more stable equilibrium state leading the network toward a certain permanent optimality of wireless network performance.

The modeling of SALOHA has been the subject of several scientific research projects in recent years. In order to be able to study the behavior of users while accessing the medium access control (MAC) layer [ZOU 54, GHA 10, ANT 14], the author in [ALT 03] presents SALOHA as being a stochastic game with imperfect information, the objective of which is to attain system equilibrium through the strategic choice of retransmission probability q_r. In [MAC 01, MAC 03], the author presents a game model of the egotistical behavior of nodes during MAC access in order to determine the right traffic load to give equilibrium in the network. Also, in [ALT 04, MA 06, YU 04, SAB 11], the authors study SALOHA in the form of a strategic game model to discuss network wireless performance, in particular the time and system output factor, as well as energy consumption. Each of these books has considered a different game model in the relationship between users and the rate of information exchanged between them. Our model belongs to the same context, insofar as we are considering a strategic erasure-coding game model, the objective of which is to observe the behavior of the network in terms of the type of coding used and to find the equilibrium point at which the protocol must be stable and high-performing in terms of output, energy savings and QoS [EL 03, PER 12, BAC 13, SAH 14, HOU 14, NIY 07].

As we saw in the previous section, SALOHA has a very high collision rate that causes successive retransmissions resulting in long transmission times. This has a negative impact on network performance, and particularly on output and energy savings. In addition to collisions, we can also see fading and noise phenomena characterizing the radio link, which cause losses and errors in packets during their transmission, leading the network into a situation of instability with completely degraded performances. In such situations, the implementation of a system of securitization and correction of transmitted data is vital to ensure a certain level of stability and reliability in wireless networks. To this end, we have integrated into our transmission system an error-detection/correction code based on Reed–Solomon (RS) codes, with a symbol-marking technique used for erasures.

3.3.2. *RS erasure codes*

RS codes are error-detection/correction codes with a Hamming distance of $d = 2*t+1$, where t is the corrective capacity. The coder codes a word of length *k* original packets into a code word $C(x)$ of N coded packets, with $N-k$ redundant packets. Upon reception, the decoder recovers the original word (*k* original packets) from the received word $C^{'}(x)$, after having detected and corrected the errors and erasures committed on the transmitted word. Thus, we define an error as being an erroneous symbol inside the code word received, whose position is unknown, while erasure is an erroneous symbol inside the received code word whose position is precisely marked. Correction consists of restoring the original value of each of the erroneous symbols in the code word received. To address erasures, we use a technique of marking each symbol, for example, determining the number of bits per symbol and the parity bit use; the latter, which is used to detect all errors whose number is uneven and to separate erasure errors in the received code word $C^{'}(x)$. In [DAN 99], the author develops an RS code (127, k, d) defined by the generator polynomial $g(x)$, the original polynomial of form $P(x) = x^7 + x^3 + 1$ where α is its root, a Galois field (2^7) containing 127 elements, and a symbol size of 7 bits. The message size is k symbols with a Hamming distance d. The number of correctible errors is t, and the maximum number of erasures is $2t$.

The generator polynomial is calculated as follows:

$$g(x) = \prod_{i=0}^{d-2} (x - \alpha^i) = (x - \alpha^0)(x - \alpha^1) \dots (x - \alpha^{d-2}) \quad [3.12]$$

Such that the word $C(x)$ is in the following form:

$$C(x) = x^{N-k} \sum_{i=k=-1}^{0} \alpha_i x^i + \sum_{j=N-k-1}^{0} r_j x^j \quad [3.13]$$

Such that $r_j \in \{0,127\}$ are the coefficients of polynomial $R(x)$, which is the remainder of the division of $X^{(N-k)} * M(x)$ by $g(x)$, where $M(x)$ is the information polynomial given as follows:

$$M(x) = \sum_{i=k=1}^{0} \alpha_i x^i = \alpha_{k-1} x^{k-1} + \ldots + \alpha_1 x^1 + \alpha_0 x^0 \quad [3.14]$$

So, the redundancy $R(x)$ is equal to:

$$R(x) = \sum_{j=N-k-1}^{0} \alpha_j x^j = r_{n-k-1} x^{N-k-1} + \ldots + r_1 x^1 + r_0 x^0 \quad [3.15]$$

The coefficients of the polynomials can be represented in the form of discrete values of between 0 and 127, or in the form of the power of α. The generation of redundancy is well explained in [MUH 04, LES 12], in which the author gives simple examples for the formation of redundant packets. For example, the first redundant packet can be generated by the function given below:

$$c_{k+1,i} = \left(\sum_{j=1}^{k} c_{j,i} \right) \bmod 2 \quad [3.16]$$

where $c_{i,j}$ is the *i*th bit of the *j*th packet.

The second redundant packet is generated by the following equation:

$$c_{k+2,i} = \begin{cases} \left(\sum_{j=1}^{i} c_{j,i+1-j}\right) \bmod 2, \ldots\ldots\ldots\ldots 1 \le i \le k+1 \\ \left(\sum_{j=1}^{k+1} c_{j,i+1-j}\right) \bmod 2, \ldots\ldots\ldots\ldots k+2 \le i \le m \\ \left(\sum_{j=1}^{k+m-i+1} c_{i-m+j,m+1-j}\right) \bmod 2, \ldots\ldots\ldots\ldots m+1 \le i \le m+k \end{cases} \qquad [3.17]$$

Now, at the reception point, the code word received will be in the form of:

$$C'(x) = [C(x) + E(x)] \bmod 2 \qquad [3.18]$$

Such that:

$$E(x) = \sum_{j=N-1}^{0} b_j x_j = b_{N-1}x_{N-1} + \ldots + b_1 x_1 + b_0 x_0 \qquad [3.19]$$

With $b_j \in \{0,127\}$. If $E(x) = 0$, then $C'(x) = C(x)$, which is the case of a transmission without errors or erasures. There is also the decoding algorithm, which is used to correct errors and replace erasures after having detected and located them. To locate erasures, the erasure locator polynomial $\sigma_e(x)$ is used:

$$\sigma_e(x) = \prod_{h=1}^{j} (1 + \alpha^{i_h} x) = (1 + \alpha^{i_1} x)(1 + \alpha^{i_2} x)\ldots(1 + \alpha^{i_j} x) \qquad [3.20]$$

where $j = \{1, 2, 3, \ldots, e'\}$. Such that, j is the erasure rank, and e' is the number of erasures calculated on reception, with $e' \le 2t$, and

$i_1, i_2, ..., i_j \in \{0,126\}$. To locate errors, we use the error locator polynomial $\beta(x)$ given as follows:

$$\sigma(x) = \sigma_e \beta(x) \quad [3.21]$$

Such that $\sigma(x)$ is the error- and erasure-detecting polynomial developed by the Berlekamp–Massey algorithm. Thus, the roots of $\beta(x)$ in the form of the power of α are used to determine the positions of the errors in $C'(x)$. The authors in [HAN 04, LEE 00, LIN 04] give a detailed explanation of the encoding and decoding process by giving concrete and simple examples of the development of error and erasure location and correction algorithms used to recover the original word.

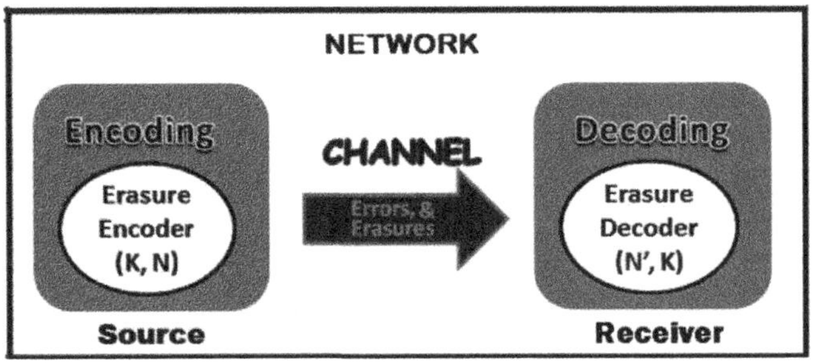

Figure 3.6. *General principle of RS erasure encoding/decoding*

A condition that must always be considered during our study, as demonstrated and verified in [SCH 90], is the following: the decoder must receive a minimum number of $N' \geq k+2$ coded packets in order to be able to extract the k original packets. In addition, RS erasure codes are characterized by an evaluation parameter called the coding (or decoding) cost Cc, which represents the coder/decoder speed given by the total number of arithmetical operations carried out during the encoding/decoding of the data to be transmitted, such that, for a coding $RS(N,k)$ the cost is equal to:

$$Cc = k\rho \log_2 N \quad [3.22]$$

where $\rho = N - k$, which represents the number of redundant packets. Thus, we can see that Cc is a function of redundancy, and it can influence Cc in a positive ($Cc_{\downarrow}$) or negative ($Cc_{\uparrow}$) way in terms of network speed. In summary, we say that the objective of erasure encoding is to recover the original word composed of k original packets via the sending of N coded packets that form the code word to be transmitted in the channel. Upon reception, the receiver receives a code word affected by errors and/or erasures, which will be composed of N' coded packets. At decoding, we can recover the k original packets from the N' packets received, as shown in Figure 3.6.

3.3.3. *The impact of erasure encoding on SALOHA*

The integration of erasure encoding introduces redundancy, and if this is well structured it will then be possible to correct serious errors experienced during transmissions. Thus, the generation of redundancy remains a very complicated part of the coding process, and constitutes the key to the success or failure of the entire transmission operation. On the one hand, the addition of additional quantities of data increases traffic flow and may cause network saturation in consequence. On the other hand, however, the rapid recovery of original data upon reception via the use of redundancy makes for a successful transmission and frees the system from having to conduct successive retransmissions, which affect wireless network performances, thus reducing traffic overload in the system. The author in [KIN 05] shows the benefit of erasure codes for the output and transmission time of SALOHA, with a random use of redundancy. We can see in Figure 3.7 that the benefit for output is limited in relatively low traffic loads, while during high traffic periods the two curves are superimposed on one another to show non-different behavior compared to the conventional protocol. This is not enough to prove the estimated contribution of erasure coding to the SALOHA protocol, and we believe that a more precise use of erasure codes may lead the network to better performances than before. To this end, we will now examine the influence of redundancy type on SALOHA network performance in the form of a strategic game.

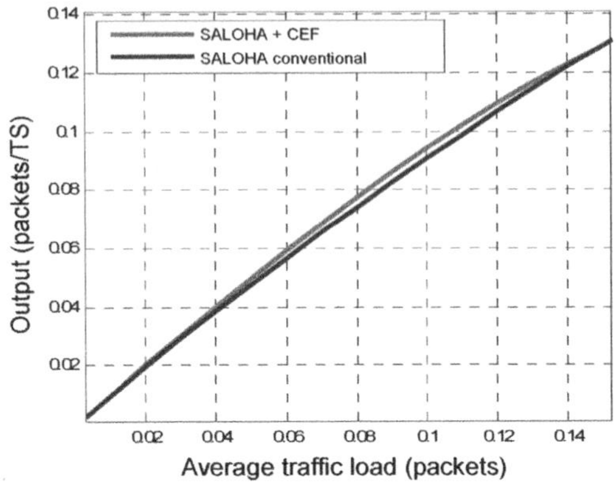

Figure 3.7. *Improvement contributed by non-strategic use of erasure codes. For a color version of this figure, see www.iste.co.uk/benslama/adhocnetworks.zip*

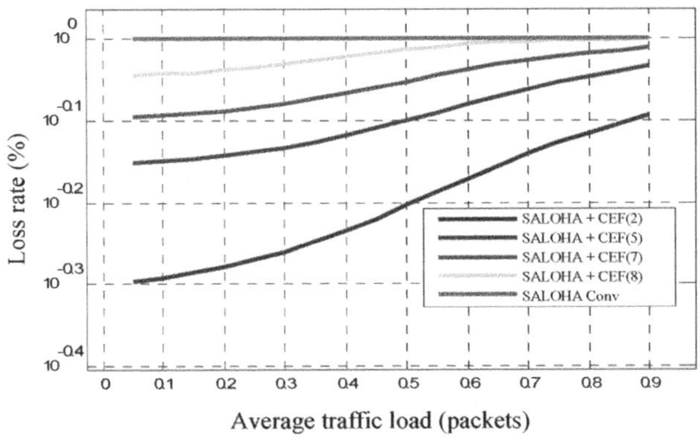

Figure 3.8. *Loss rate for coding of packets transmitted depending on redundancy type. For a color version of this figure, see www.iste.co.uk/benslama/adhocnetworks.zip*

3.3.4. *Description of game model*

This is a wireless network composed of M nodes sharing a transmission channel of limited bandwidth and using a random access algorithm, SALOHA, to access the medium. We will assume that the nodes are equipped with erasure coders and decoders, and during each transmission a coder sends its code word with a different redundancy from other transmissions of a new word, and which is also different from those used by the other coders (nodes). Before starting transmission, the source node i, $i = 1,...,M$ codes its original word (composed of k original packets) into a code word of N coded packets ($N > k$) with redundancy ρ_i and written as CEF(ρ_i). Upon reception, the node receives a code word composed of N' packets, coded so that ($N' \neq N$). The receiving station extracts the k original packets through the N' packets received using the error and erasure correction and replacement mechanisms available to the erasure decoder.

During the progression phase, the node transmits its data with a redundancy that also guarantees the recovery of all original packets and also maximizes transmission output. Thus, when a node uses a sufficiently large redundancy ρ_i this increases the traffic load elsewhere, and if the transmission channel is free enough to accommodate the packets sent, there will be a higher probability of success, which will then be manifested by a considerable increase in transmission output Th_i. Here, we must take into consideration the idea that the other nodes forming the network also think (and behave) in the same way as the first node, with performance maximization being the objective of each of them, such that node i', $i' = 1,...,i-1,i+1,...,M$ sends its word with a redundancy $\rho_{i'} \neq \rho_i$, and thus CEF($\rho_{i'}$) to have a relative output $Th_{i'}$. Based on this idea, the network will find itself faced with an enormous traffic load, with a large number of nodes wishing to transmit. In this case, the number of free slots will be smaller, and for the node to complete its transmissions it must preferably reduce its redundancy to the minimum actionable, both in order to be able to transmit all its packets in the remaining free slots that exist, and to avoid network saturation. The operating process of the nodes during the two transitory states of the SALOHA system is detailed in Figure 3.9. We can conclude, then, that in order to keep the network in a stable and optimal state of equilibrium for a more prolonged period of time, we must manage the nodes' behavior (strategy) with regard to the quantity of redundancy used, and also with regard to the state of operation of the SALOHA network. To this end, we have formalized all of this in a strategic game model based on game theory.

The nodes in our model take the role of *players*, such that a player i enters the game when there are packets to transmit; it is then called an active player, and it leaves the game once it has successfully transmitted all of its packets. We assume that there is no cooperation and no exchange of information between the players (real case), and that each player i chooses the strategy s_i that optimizes its profit, and that after each transmission failure the player repeats its strategies until all its transmissions have been successful. Here, the player's *strategy* consists of choosing the best redundancy ρ_i, which maximizes its *utility* u_i and is expressed in terms of output Th_i, minimizing transmission time as well. Thus, we have a *repeated non-cooperative game with imperfect information*. Figure 3.9 shows an example of the game model of a three-player (node) network in which players compete to win free slots in order to complete their transmissions.

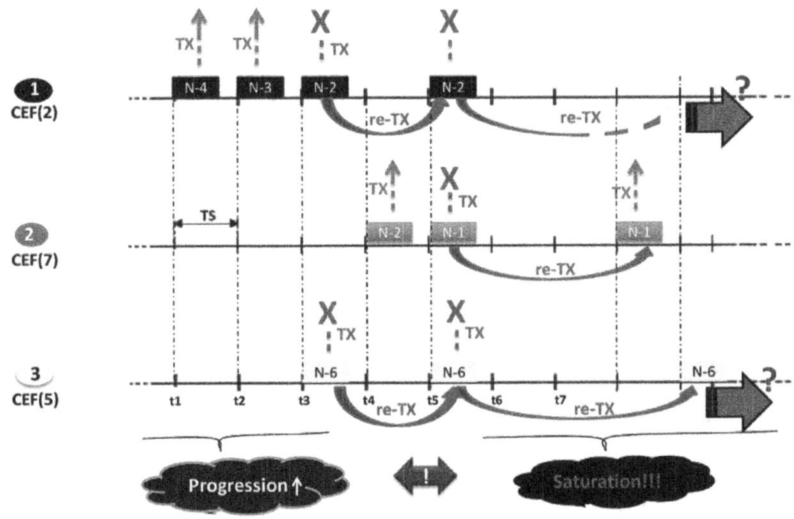

Figure 3.9. *Intermittence of SALOHA network by adding redundancy. For a color version of this figure, see www.iste.co.uk/benslama/adhocnetworks.zip*

Game J is triple; such that $J = (\{1,...,M\}, s, u)$ where:

– $\{1,...,M\}$ represents all of the players in the game.

– s represents all of the players' strategies, such that s_i is the strategy of player i, and s_{-i} is the strategy of all the players except player i. Thus, $s = \{s_1, s_2, ..., s_M\} = \{\rho_1, \rho_2, ..., \rho_M\}$.

– u represents all of the players' utilities, such that u_i is the utility of player i.

3.3.5. *Study of utility*

Utility in our game is transmission output, and to evaluate this we use the same probability concept used in [KIN 05], taking into consideration the number of active players in the network as well as the type of redundancy used. Transmitted packets form a Poisson process with an average value λ. After coding, the average traffic load becomes $\lambda.N/k$. The system's output is thus expressed by the following expression:

$$Th = \lambda.Ps \qquad [3.23]$$

where:

$$Ps = P_k + \sum\nolimits_{n'=1}^{k-1} \sum\nolimits_{m=1}^{n'} m/k \, P_{n',m} \qquad [3.24]$$

The probability that a player $i \in M$, which is not on hold, will transmit its packet using a given slot is equal to:

$$P_\rho = (e^{-\lambda(1+\frac{\rho}{k})})(1-e^{-\lambda(1+\frac{\rho}{k})})^{M-1} \qquad [3.25]$$

Such that $\frac{N}{k} = 1 + \frac{\rho}{k}$. We write $Q_\rho = 1 - p_\rho$.

The probability that at least k coded packets will be successfully received is:

$$P_k = \sum_{i=k}^{N} \binom{N}{i} P_\rho Q_\rho^{N-i} \qquad [3.26]$$

The probability that only $n'(n' \leq k)$ coded packets will be received, when m of n packets are original, is:

$$P_{n',m} = \binom{k}{m}\binom{N-k}{n'-m} P_\rho^{n'} Q_\rho^{N-n'} \quad [3.27]$$

In fact, an original packet can be successfully recovered if at least k of N coded packets are successfully received, or if $n'(n' \leq k)$ coded packets are received, when m of n' packets are original. Thus, we write:

$$Ps = \sum_{i=k}^{N}\binom{N}{i} P_\rho Q_\rho^{N-i} + \sum_{n'=1}^{k-1}\sum_{m=1}^{n'}\frac{m}{k}\binom{k}{m}\binom{N-k}{n'-m} P_\rho^{n'} Q_\rho^{N-n'} \quad [3.28]$$

According to equation [3.23], the expression of utility is:

$$u(\rho) = Th(\rho) = \lambda.\left\{\sum_{i=k}^{N}\binom{N}{i} P_\rho Q_\rho^{N-i} + \sum_{n'=1}^{k-1}\sum_{m=1}^{n'}\frac{m}{k}\binom{k}{m}\binom{N-k}{n'-m} P_\rho^{n'} Q_\rho^{N-n'}\right\} \quad [3.29]$$

3.3.6. *Discussion of equilibrium*

To discuss game equilibrium, we refer to all of the theorems presented in the first part of Chapter 2, first to demonstrate the existence or non-existence of game equilibrium, and then to find its value in the case of existence and thus prove whether it is unique or not.

3.3.6.1. *Existence*

As we can see in equation [3.29], the utility of the game is a function of three variables, λ, M and ρ. So, we take the other two variables as a constant value in order to examine only the effect of ρ on the utility $u(\rho)$. Thus, we choose passive values that do not affect the overall value of the function, and for this we write $\lambda = 1$ and $M = 1$ for a single network user. The curve in Figure 3.10 describes the output behavior in terms of redundancy, and we can see that the function is continuous and concave on interval $[0,10]$, such that $\frac{\partial^2 Th_i}{\partial \rho_i^2} < 0$. Thus, according to theorem 2.1 of *Weierstrass*, cited in section 2.3.2; a maximum s_i^* exists such that for $\forall s_i \in s, \exists s_i^* \in s$ such that $u(s_i^*, s_i) > u(s_i^{'}, s_i)$ where s_i^* is the dominant strategy. And according to theorem 2.2of *Rosen* in section 2.3.2; our utility possesses a unique equilibrium. Theorem 2.4 of *(Debreu, Glicksberg and Fan)* in section 2.3.2 confirms that our non-cooperative game admits a unique NE in pure s_i strategies, such that the sets of strategies s_i are closed and limited, and also convex when combinations of strategies belong to the compact set s themselves. Thus, we conclude that a pure and unique *NE* exists for our non-cooperative imperfect-information coding game $J = (\{1,...,M\}, s, u)$.

3.3.6.2. *Evaluation*

Now, to find the NE, we analyze the best response of players i. In Figure 3.10, we can see that, for $\rho = 0$, the network presents a very low output, which increases progressively up to a value of $\rho = 2$, where it reaches its maximum value for all the nodes in the network, such that $Th(\rho = 2) = \arg\max Th$, and we mark this optimal point by ρ^*

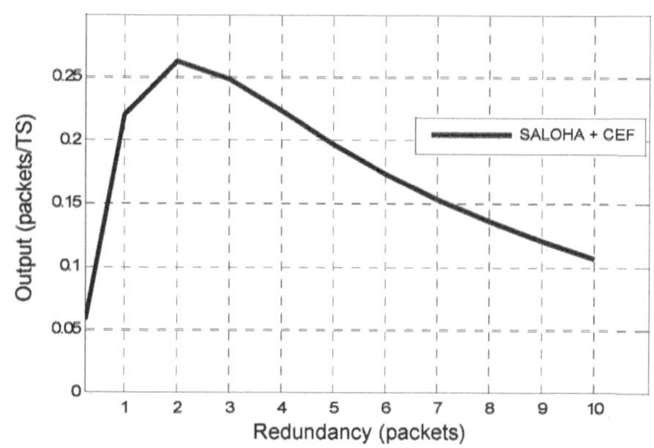

Figure 3.10. *Variation in network output (SALOHA + CEF) depending on redundancy*

Afterward, when $\rho > 2$, the network degrades gradually toward lower output levels. Thus, we can write that when $\rho < \rho^*$ congestion dominates and causes low network output. However, when $\rho > \rho^*$ output drops toward lower levels. Thus, strategy $Th(\rho^*, \rho_{-i})$ presents the dominant strategy, such that $Th(\rho^*, \rho_{-i}) \geq Th(\rho', \rho_{-i})$, for any $\rho' \neq \rho^*$, and it also presents the best response of player i to strategies ρ_{-i}; and we write $\rho^* = BR(\rho_{-i})$. According to definition 2.1 and definition 2.2 of section 2.2.6.2, we conclude that strategy $Th(\rho^*, \rho_{-i})$ is the pure and unique NE of our coding game. Next, we will demonstrate the benefit contributed by equilibrium to the functioning of SALOHA in an *ad hoc* network.

3.4. SALOHA network performance at Nash equilibrium

To evaluate the difference in terms of performance between the network at equilibrium and the network outside of equilibrium, we will begin by examining coding cost to see the impact of equilibrium on the functioning of RS erasure coders/decoders, and then we will look, respectively, at packet loss rate, output, stability, transmission time and finally energy management.

3.4.1. *Coding cost*

Figure 3.11 shows that during equilibrium, coding/decoding operations present the lowest coding/decoding cost. Consequently, coders/decoders function more quickly at equilibrium than elsewhere, and the difference in speed contributed in terms of coding/decoding results in a remarkable increase in system output, when the node is able to take on a larger amount of data (flow) to transmit, such that during equilibrium, the coding/decoding cost varies between $[0.9, 2.2\%]$, which means that the coder/decoder executes 90 mathematical operations to generate two redundant packets or to recover eight original packets. To generate 10 redundant packets, the coder executes up to 1,000 mathematical operations. We can see a significant difference, which can have an enormous impact on data transmission output.

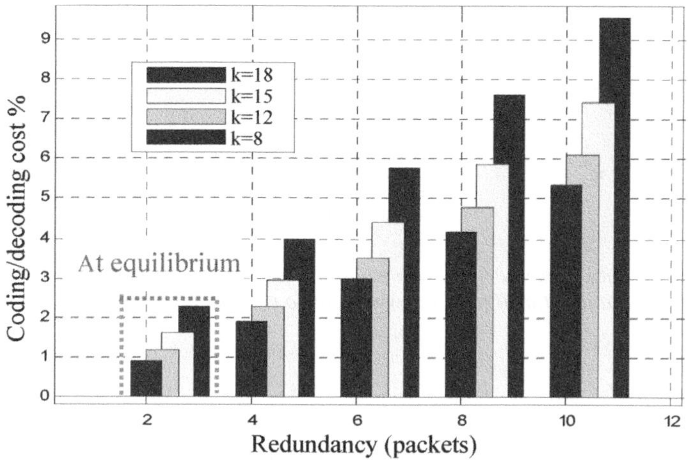

Figure 3.11. *Coding/decoding cost in terms of redundancy and packet size. For a color version of this figure, see www.iste.co.uk/benslama/adhocnetworks.zip*

In fact, the difference will be even greater in the real case, with a number on the order of 10,000 mathematical operations to generate or recover a larger number of coded packets. Thus, at equilibrium the system functions more rapidly compared to a conventional system, when the RS condition necessary to recover k original packets (section 3.2.2) is always respected.

3.4.2. *Loss rate*

Moreover, the advantage of NE can also be seen in the loss rate of packets transmitted. At equilibrium, the network presents a much lower loss rate than the one seen outside equilibrium, and Figure 3.13 shows a considerable discrepancy between the two systems. This observed advantage also confirms the increased successful packet transmission rate at equilibrium.

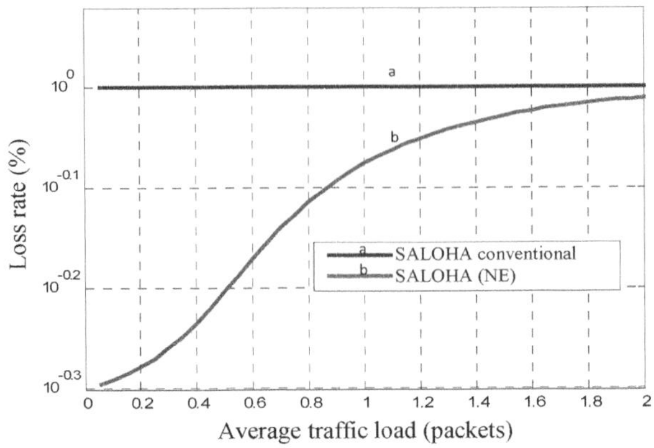

Figure 3.12. *Packet loss rate at Nash equilibrium and outside Nash equilibrium. For a color version of this figure, see www.iste.co.uk/benslama/adhocnetworks.zip*

3.4.3. *Output*

In Figure 3.13, we can see the influence of the number of users on system output; output varies in an inversely proportional manner to the number of users in transmission, such that when the number of users increases, output gradually drops until the network's saturation point is reached, when it becomes almost nil. We can also see that moving from 5 to 10 users causes a sharper drop in output than the other transitions from 10 to 15 users and from 15 to 20 users. This can be explained by the fact that the network is entering the saturation zone (marked by the 5–10 transitions) and undergoes a considerable degradation in transmission output. Next, and once it has entered this saturation zone, the network deteriorates progressively by negligible degrees until it reaches the limiting point of saturation, where output becomes almost nil (system stoppage).

Figures 3.14(a) and (b) show how SALOHA functions at equilibrium with a higher output in comparison to its functioning in the classic state. This superiority becomes even greater when a larger number of users are participating; in fact, with a single user and thus an absence of collisions, the network shows a higher output in comparison with a 20 user network where the number of collisions is higher.

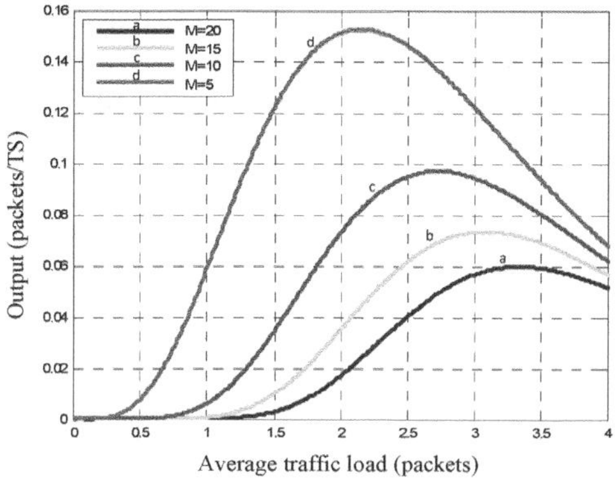

Figure 3.13. *SALOHA output with different numbers of users. For a color version of this figure, see www.iste.co.uk/benslama/adhocnetworks.zip*

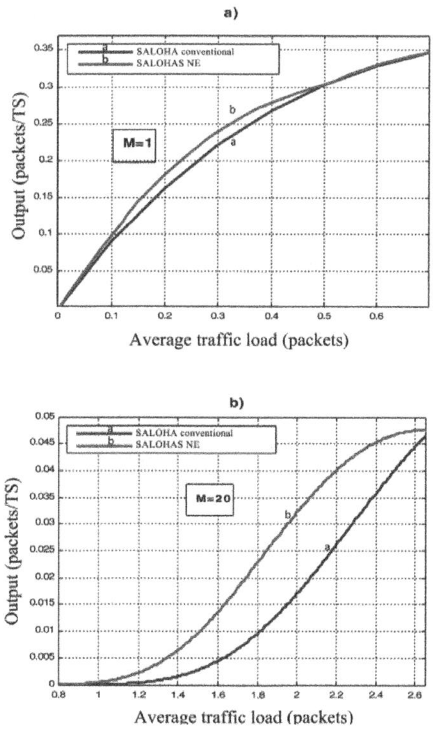

Figure 3.14. *a) SALOHA network with a single user. b) SALOHA network with multiple users. For a color version of this figure, see www.iste.co.uk/benslama/adhocnetworks.zip*

Again, the benefit for SALOHA network performance, particularly in terms of transmission output, can be clearly seen in the two figures, and especially in the second figure, which shows a more practical case. Thus, while approaching reality, the benefit of NE becomes even more marked and significant, which also confirms the usefulness of game theory in the analysis and solution of issues like these.

3.4.4. *Stability*

In addition, system stability at equilibrium is reinforced further when the zone, where output is higher, is enlarged again to offer the system an expanded margin of optimization. To confirm this, we return to the accumulation recorded between two successive states of the Markov chain, such that the variation of $A(n)$ it expresses accumulates packets on hold, which can give us an idea of the stability of the SALOHA protocol, particularly in the event of a very high traffic load. Thus, according to equation [3.6], we can write:

$$\lim_{\lambda \to \infty} A(n) = \lim_{\lambda \to \infty} (M-n)q_t - P_{succ}. \qquad [3.30]$$

Such that $q_t = 1 - e^{-\frac{\lambda}{M}}$. So $\lim_{\lambda \to \infty} q_t = 1$ with $\lim_{\lambda \to \infty} e^{-\frac{\lambda}{M}} = 0$.

Also:

$$\lim_{\lambda \to \infty} P_{succ} = \lim_{\lambda \to \infty} Ps = \lim_{\lambda \to \infty} \sum_{i=k}^{N} \binom{N}{i} P_\rho Q_\rho^{N-i} + \sum_{n'=1}^{k-1} \sum_{m=1}^{n'} \frac{m}{k} \binom{k}{m} \binom{N-k}{n'-m} P_\rho^{n'} Q_\rho^{N-n'}$$

Such that $\lim_{\lambda \to \infty} P_\rho = \lim_{\lambda \to \infty} (e^{-\lambda(1+\frac{\rho}{k})})(1-e^{-\lambda(1+\frac{\rho}{k})})^{M-1} = 0$. Thus, we have $\lim_{\lambda \to \infty} P_{succ} = 0$.

So:

$$\lim_{\lambda \to \infty} A(n) = (M-n) = c \qquad [3.31]$$

When all the nodes of the SALOHA network are at equilibrium, each of them maximizes its probability of successful transmission, and consequently the number of holding packets n will be lower, with a slight variation among the different system states. Thus, the accumulation limit, which is equal to c, tends to be almost fixed (slight variation) when the traffic load tends toward infinity (theoretically). Despite this, this result shows the consequences of SALOHA saturation in the event of a very high traffic load, but it also shows the network's stability during equilibrium.

3.4.5. *Transmission time*

At equilibrium, SALOHA offers shorter transmission time than is conventional outside equilibrium. Figure 3.15 shows a considerable advantage in transmission time, contributed by NE in the SALOHA network. Despite this, the time always shows a sudden variation toward higher values in the event of very low traffic loads. This disadvantage remains one of the major challenges which the scientific researchers currently facing.

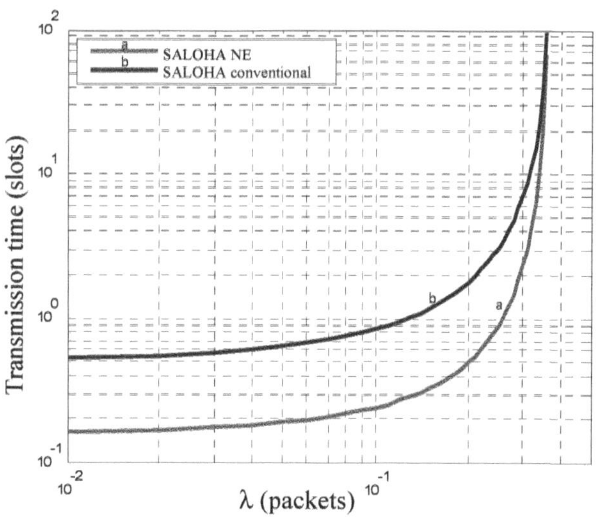

Figure 3.15. *Transmission time in SALOHA in NE and outside NE. For a color version of this figure, see www.iste.co.uk/benslama/adhocnetworks.zip*

3.5. Conclusion

The coding game study proposed in this chapter has enabled us to examine NE in the SALOHA network. At equilibrium, the network becomes higher performing in terms of loss rate, transmission time and transmission output, as well as energy consumption. In order to assess the benefits of equilibrium for SALOHA networks, we have presented simulations containing comparisons of different qualities such as coding/decoding cost, packet loss rate, transmission time and output. Simulations confirm the advances contributed by equilibrium to the various performances of a SALOHA network [ANS 14].

4

Games in CSMA Networks

4.1. Introduction

Though the Slotted ALOHA (SALOHA) protocol is used much more widely in satellite-terrestrial station links to make reservations when the allocation of bandwidth is requested, enhanced distributed coordination function (EDCF) is currently used heavily in *ad hoc* networks, and particularly in mobile *ad hoc* networks (MANETs) to control access to transmission support in the medium access control (MAC) layer. In fact, EDCF was originally a developed version of the Carrier Sense Multiple Access/CSMA with Collision Avoidance (CSMA/CA) protocol, which is part of the family of random access protocols, and, to access the medium, nodes in *ad hoc* mobile networks use the CSMA/CA random access technique to avoid collisions during data transmission. This protocol benefits from several control and acknowledgment mechanisms in order to better control access to the medium and ensure a certain level of optimality in transmission output and energy management. These mechanisms include request to send (RTS), clear to send (CTS), acknowledgment (ACK) and back off, which have already been explained in detail in Chapter 1. Despite all these prevention and control techniques, CSMA and its improved versions still suffer from collisions and congestion, which have a negative effect on their performance and reduce their effectiveness at successfully transmitting data and saving energy. Researchers continue to suggest alternative solutions for this communication protocol, including those in references [BAE 12,

CHA 04, ZHA 00, WAN 05]. It is clear, though, that the various solutions proposed in the literature to improve the performance of CSMA networks develop some aspects of performance to the detriment of others; for example, solutions that seek on one hand to benefit network capacity and energy control may also reduce the network's global output and quality of service (QoS). Unfortunately, though some problems have been removed, these solutions are based on certain techniques in which the analysis of phenomena and their causes is made complicated by the pileup of protocols and their various interactions. Thus, it is necessary to seek out flexible and intelligent solutions that can satisfy all of the aspects of performance of the *ad hoc* network while respecting the basic architecture of the network, in order to be able to guarantee a reliable and adequate level of QoS for real time and multimedia applications. In recent years, the analysis of CSMA via game theory has been the subject of several scientific projects that have provided new and effective suggestions for the yield of this random protocol; for example, in references [WAN 05, JIA 09, CAG 05, KON 06, QUE 04, QUE 05, YAN 10, KLE 75, AKK 11, GHA 13b], the authors study the egotistical behaviors of nodes in CSMA/CA networks through a non-cooperative repeated game, and propose solutions based on Pareto-optimal Nash equilibrium (NE) [GAS 13, GHA 13a, SHA 13, HAR 99, BAS 94] to optimize the network. Additionally, in [ANT 10, ELI 11, ZHA 09, MAS 09], the authors examine the possibility of cooperating with network users in order to maximize output via a repeated cooperative game. To this end, and in the same context, in this chapter we will offer practical solutions to reduce collisions and correct errors in order to maintain the CSMA network at its optimum level for a longer period of time and with more stability.

This chapter is structured into two main parts. In the first part, we will discuss the state of CSMA performance in conventional mode, where we examine transmission delay, packet loss rate and transmission throughput in order to explain the low level caraterizes the performances of CSMA network. However, in order to plan for more effective solutions, we will begin by examining the problems experienced by this protocol and explaining the causes of the degradation of CSMA performance, as well as the sources of these causes. As we will see, long frames, hidden nodes and exposed nodes are the principal causes of the collision and successive retransmission phenomena that prolong transmission times and reduce output by leading the network to dry losses of available energy. To solve these problems, we will propose, as with SALOHA, two complementary solutions; the first solution consists of integrating an erasure code in order to recover all original packets upon reception and thus increasing the probability of successful transmission and reducing the number of retransmissions. The second solution is based on game theory; we will propose a mathematical model of a strategic coding game that will allow us to study the egotistical behavior of nodes during access to the medium and to see the possibility of converging toward a beneficial solution that will satisfy all of the nodes and put the network in a more stable state of equilibrium. We will begin by introducing the game model by defining its vital components and describing its various characteristics. Next, we will move on to the most difficult stage in game theory: the study of equilibrium. Here, we will discuss the existence of equilibrium and the possibility of convergence for our model. As we will show, our model converges toward a unique solution that represents pure and unique NE for our non-cooperative, imperfect-information coding game. To assess the impact of our proposed solutions on CSMA network performance, in the second part of this chapter we will give a set of simulations describing the various performances of the network at equilibrium, and compare them with performances in conventional mode. Finally, we will look at the benefits provided by our model to the CSMA network.

4.2. CMSA performance

In an *ad hoc* network, a node can either be in transmission mode or in reception mode, and shifting between these two modes is done in a very short amount of time that is negligible compared to the propagation time. The same thing is true for the carrier detection time, which is brief in practice and can be disregarded when calculating overall transmission time. In addition, as in SALOHA, we will assume points a, b, c, e, f and g from section 3.1 of Chapter 3. The CSMA protocol proposed in [TOB 80] offers several variants including persistent CSMA, p-persistent CSMA and non-persistent CSMA. Since the latter offers the best performance in terms of transmission output and time, we will consider non-persistent discretized CSMA in all of the calculations and analyses presented in the rest of this chapter.

To limit interference levels, in CSMA the node wishing to transmit listens to the transmission channel first, and if it finds that the channel is free it transmits the packet; otherwise, it postpones retransmission to after a few time slots (TS) depending on propagation time. Before retransmission, the node listens to the channel again and goes back to the first stage of the algorithm. In discretized CSMA, the time axis is divided into TS, such that all transmitters are synchronized to begin their transmissions at the start of each TS. If T is the propagation time (in seconds) of a packet and includes the time of transmission detection in the channel, and α is the ratio between the propagation time and transmission time of a packet, then $\alpha = \tau/T$, such that τ is the time necessary to transmit a packet. We also say that α is the CSMA period of vulnerability, such that, in a time axis t_0 is the arrival time of the packet that has heard an open channel. So, if another packet arrives during time $(t_0, t_0 + a)$ when it finds the channel free, its transmission will have a collision. Otherwise, the first packet to have arrived will be successfully transmitted.

To analyze the CSMA protocol, we can use the Markov chain presented in Chapter 3 (Figure 3.3), in which the system state is a stochastic process with discrete time and in discrete state n, which represents the number of packets on hold, with the probability of transition by the system from state n to state $n+i$ given by transition matrix [3.4]. The time between two successive transitions is equal to α in the event of a free slot, and to $1+\alpha$ in the event of an occupied slot directly followed by a free slot. Thus, during the transition to state n, the probability that there will be no transmissions in the next slot is:

$$P_{SL} = e^{-\lambda\alpha} q_t^n \qquad [4.1]$$

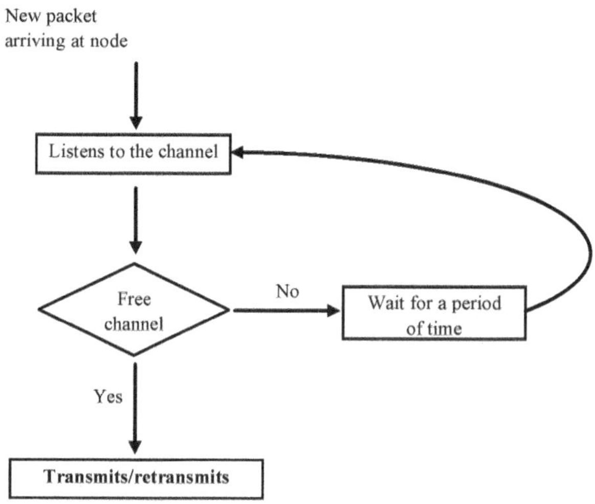

Figure 4.1. *Basic CSMA algorithm*

Such that P_{SL} is the probability of a free slot. Thus, the time elapsed between the two transitions is equal to:

$$D_{SL} = \alpha + 1 - e^{-\lambda\alpha} q_t^n \qquad [4.2]$$

And the estimated number of packets arriving between transitions in state n is:

$$E\{\text{packets arriving}\} = \lambda\left[\alpha + 1 - e^{-\lambda\alpha} q_t^n\right] \qquad [4.3]$$

In fact, the number of packets arriving during transitions is nearly the same as the probability of successful transmission, so we write:

$$P_{succ} = \left(\lambda\alpha + \frac{q_r n}{q_t}\right) e^{-\lambda\alpha} q_t^n \qquad [4.4]$$

With $q_r < 1$. And, accumulation in state n will be equal to:

$$A_n = \lambda(\alpha + 1 - e^{-g(n)}) - g(n)e^{-g(n)} \qquad [4.5]$$

where, $g(n) > \lambda$ is the total number of transmissions following the transition to state n:

$$g(n) = \lambda\alpha + q_r n \qquad [4.6]$$

Output is given by the following equation:

$$Th = \frac{\alpha g(n) e^{-\alpha g(n)}}{(1 - e^{-\alpha g(n)}) + \alpha} \qquad [4.7]$$

Such that the numerator of equation [4.7] represents the number of packets transmitted per transmission, while the denominator is the estimated time of transition between two successive states.

Though it shows remarkable progress in terms of output compared to SALOHA, CSMA remains weak and unstable when faced with relatively low traffic loads, and its margin of optimization appears to be further reduced when the network shifts very quickly toward saturation. We can see that the choice of parameter α is very important for the improvement of CSMA, since when $\alpha \ll$ is small, output will be higher.

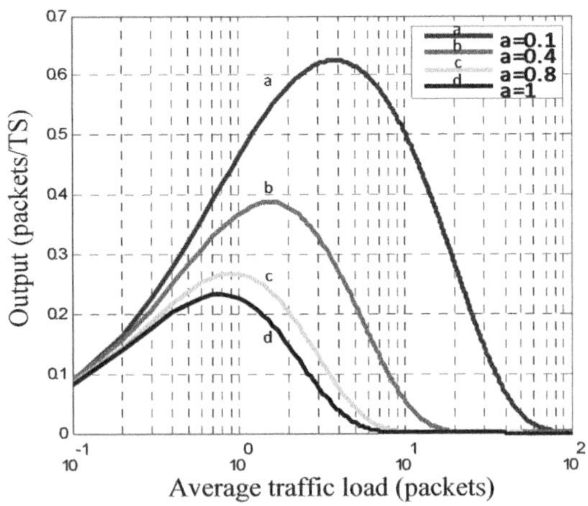

Figure 4.2. *Non-persistent CSMA with different values of* α*. For a color version of this figure, see www.iste.co.uk/banslama/adhocnetworks.zip*

4.3. Sources of problems in CSMA networks

The operating principle of CSMA is explained in Chapter 1, with definitions of all the mechanisms included in this protocol to counteract phenomena of collision and saturation. However, CSMA and its improved versions still suffer from problems of collision and congestion, making them incapable of fulfilling current needs in multimedia and real time. Also, despite its mechanism of listening to the channel before starting transmission in a wireless network, CSMA must still deal with new sources of collisions that listening to the channel cannot completely control, such that:

– In a wireless network, it is impossible to be sure of that all the nodes are listening to each other, and the fact that a node wishing to transmit tests whether the support is free does not necessarily mean that the support is free around the receiver.

– In the event of a *long frame*, the node executes multiple transmissions in order to complete the sending of the whole frame, which increases the probability of having collisions. In addition, failure during the transmission of a long frame obligates the node to start the transmission over from the beginning while respecting the CSMA mechanism, which introduces significant losses in time and cost.

– The problem of *exposed nodes*: it occurs that during transmission between two nodes in the network, a third node detects the occupied channel, which prevents it even from transmitting its packets to a node other than the two that are communicating with each other.

– The problem of *hidden nodes*: this problem occurs when two nodes cannot hear each other because the distance separating them is too large, or an obstacle is preventing them from communicating with each other, but they have coverage zones that interconnect. So, it happens that these two nodes start transmission at the same time and to the same destination. This means that there will be a collision of packets, and the receiving node in this case cannot receive any communications. Even with the RTS/CTS mechanism integrated in CSMA to avoid collisions caused by hidden nodes, it has been shown in practice that there are situations in which this mechanism is not very effective, and may even be an additional source of collision itself, following the failure to transmit its packets. Figure 4.3 shows a practical case that may exist, in which the RTS/CTS mechanism fails in the reservation of a channel. A case like this cannot be duplicated in simulators, which assume a perfectly circular carrier detection field with a radius that is twice as wide as the communication zone.

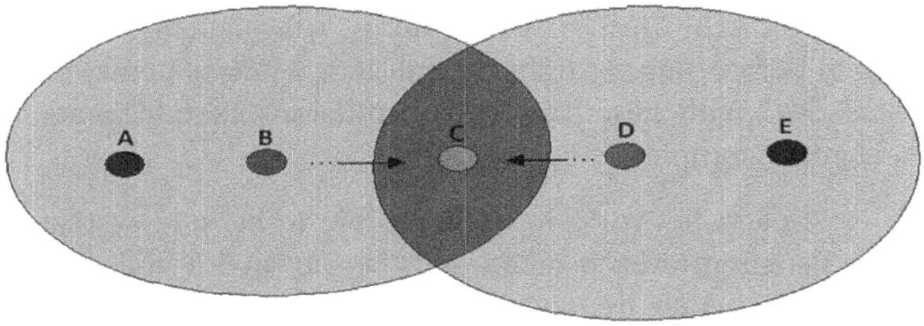

Figure 4.3. *Collision by hidden nodes (chain distribution)*

At a given time, nodes *B* and *D*, which cannot hear each other at all, wish to transmit simultaneously, which may cause a collision at node *C*. In reality, we might encounter several identical distributions and more complex ones as well, where numerous collision situations can arise. However, despite all these precautions, the CSMA network still undergoes collisions and other disruptive factors that are directly harmful to its performances. It is within this context that we are proposing a game model that formalizes conflict situations between active nodes and offers the network an equilibrium situation that optimizes its performance and makes it more efficient.

4.4. Modeling of node behavior in CSMA using a strategic coding game

4.4.1. *Game model analysis*

As in the case of the SALOHA protocol, the coding game in CSMA is based on the same principle, with the added consideration of the carrier listening function during calculation of the utility function and the evaluation of the overall transmission time. The description of the game model is given in detail in section 3.2.4 of Chapter 3. In our model, nodes act as *players*. A player i enters the game when it has packets to transmit; it is then called an active player, and it will leave the game after having successfully transmitted all its packets. We assume that there is no cooperation and no exchange of information between players (real case), and that each player i chooses the strategy s_i that optimizes its profit, and after each transmission failure the player repeats its strategies until all its transmissions have succeeded. Here, the player's *strategy* is to choose the best redundancy ρ_i to maximize its *utility* u_i, which is expressed in terms of output Th_i, and also to minimize transmission time. Thus, we are playing a *repeated non-cooperative game with imperfect information.* Figure 3.9 shows an example of the game model for a three-player (node) network in which players are competing to win free slots to complete their transmissions.

The game J is, therefore, triple, such that $J=(\{1,...,M\},s,u)$ where:

– $\{1,...,M\}$ represents all of the players in the game.

– s represents all of the players' strategies, such that s_i is the strategy of player i, and s_{-i} is the strategy of all players except player i. Thus, $s=\{s_1,s_2,...,s_M\}=\{\rho_1,\rho_2,...,\rho_M\}$.

– u represents the players' utilities, such that u_i is the utility of player i.

4.4.2. *Utility function*

The arrival of packets forms a Poisson process with an average value λ when all of the packets are of equal length. Through Figure 4.2, we choose the smallest value of α, which gives the best output in conventional mode. To extract the utility function, we must express the estimated value of the period during which the channel is occupied $\overline{B}$, and the period during which the channel remains free $\overline{I}$, with the sum of both periods forming the estimated length of the cycle. We also determine the value of $\bar{U}$, which represents the period of time within the cycle during which the network remains collision-free. After the integration of erasure coding, the probability that a player will successfully transmit its packet is:

$$P_\rho = (e^{-\lambda(1+\frac{\rho}{k})})(1-e^{-\lambda(1+\frac{\rho}{k})})^{M-1} \quad [4.8]$$

To simplify the calculations, we write; $Q_\rho = 1 - p_\rho$.

The probability that at least k coded packets will be successfully received is:

$$P_k = \sum_{i=k}^{N} \binom{N}{i} P_\rho Q_\rho^{N-i} \quad [4.9]$$

The probability that only $n'(n' \leq k)$ coded packets will be received when m of n packets are original is:

$$P_{n',m} = \binom{k}{m} \binom{N-k}{n'-m} P_\rho^{n'} Q_\rho^{N-n'} \quad [4.10]$$

In fact, an original packet can be successfully recovered if at least k of N coded packets are successfully received, or if $n(n \leq k)$ coded packets are received, when m of n packets are original. Thus, we write:

$$Ps = \sum_{i=k}^{N} \binom{N}{i} P_\rho Q_\rho^{N-i} + \sum_{n'=1}^{k-1} \sum_{m=1}^{n'} \frac{m}{k} \binom{k}{m} \binom{N-k}{n'-m} P_\rho^{n'} Q_\rho^{N-n'} \quad [4.11]$$

Now, we have the expression of output, which is given in [KRA 98] by the equation:

$$u(\rho) = \frac{\bar{U}}{\bar{B} + \bar{I}} \quad [4.12]$$

where $\overline{U} = Us.Ps.T$, $\overline{B} = T + \tau$ and $\overline{I} = \dfrac{\tau e^{-\alpha G}}{1-e^{-\alpha G}}$.

$$Us = \frac{P\{\text{only one arrival during interval } \tau\}}{P\{\text{several arrivals occur}\}} = \frac{e^{-\alpha G}(1-e^{-\alpha G})^{M-1}}{e^{-\alpha GM}} \quad [4.13]$$

where $G = \lambda N / k$. Thus,

$$Us = e^{-\alpha G(1-M)}(1-e^{-\alpha G})^{M-1} \quad [4.14]$$

This gives us:

$$u(\rho) = \frac{(1-e^{-\alpha G})^{M}.Ps.e^{-\alpha G(1-M)}}{1+\alpha-e^{-\alpha G}} \quad [4.15]$$

or, more explicitly:

$$u(\rho) = \frac{(1-e^{-\alpha\lambda(1+\rho/k)})^{M}.Ps.e^{-\alpha\lambda(1+\rho/k)(1-M)}}{1+\alpha-e^{-\alpha\lambda(1+\rho/k)}} \quad [4.16]$$

Thus, relative to an active node, we have:

$$u(\rho_i) = \frac{(1-e^{-\alpha\lambda(1+\rho_i/k)})^{M}.Ps_i.e^{-\alpha\lambda(1+\rho_i/k)(1-M)}}{1+\alpha-e^{-\alpha\lambda(1+\rho_i/k)}} \quad [4.17]$$

4.4.3. *Discussion of equilibrium*

We refer to all of the theorems presented in the first part of Chapter 2, first to demonstrate the existence or non-existence of game equilibrium, and then to find its value in the case of existence and thus prove whether it is unique or not. We can see that the utility function is represented by the curve in Figure 4.4, which is continuous and concave on interval $[0,20]$. Thus, according to theorem 2.1 of *Weierstrass*, cited in section 2.2.2; a maximum s_i^* exists such that for $\forall s_i \in s$, $\exists s_i^* \in s$ such that $u(s_i^*, s_i) > u(s_i', s_i)$ where s_i^* is the dominant strategy. And according to theorem 2.2 of *Rosen* in section 2.2.2; our utility possesses a unique equilibrium. Theorem 2.4 of (*Debreu, Glicksberg and Fan*) in section 2.2.1 confirms that our non-cooperative game admits a unique NE in pure s_i strategies, such that the sets of strategies s_i are compact, and also convex when combinations between them also belong to the compact set s_i themselves. Thus, we conclude that a pure and unique *NE* exists for our non-cooperative imperfect-information coding game $J = (\{1,...,M\}, s, u)$. Now, to find NE, we analyze the best response of players i.

In Figure 4.4, we can see that, for $\rho = 0$, the network presents a very low output, which increases progressively up to a value of $\rho = 5$, where it reaches its maximum value for all the nodes in the network, such that $Th(\rho = 5) = \arg\max Th$, and we mark this optimal point by ρ^*. Subsequently, when $\rho > 5$ the network degrades gradually toward lower output levels. Thus, we can write that when $\rho < \rho^*$ congestion dominates and causes low network output. However, when $\rho < \rho^*$ output drops toward lower levels.

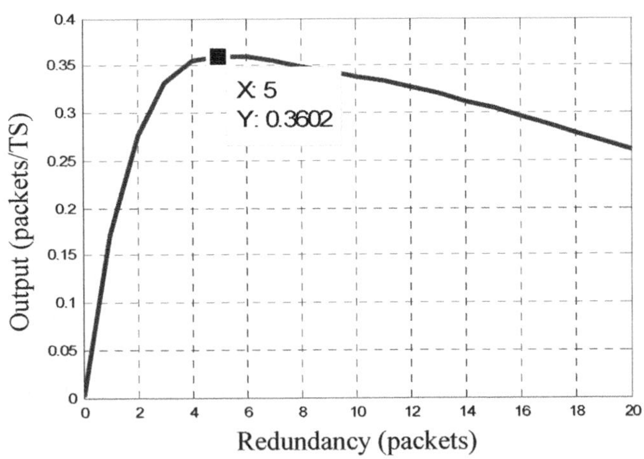

Figure 4.4. *Variation of output in relation to redundancy*

Thus, strategy $Th(\rho^*,\rho_{-i})$ presents the dominant strategy, such that $Th(\rho^*,\rho_{-i}) \geq Th(\rho',\rho_{-i})$, for any $\rho' \neq \rho^*$, and it also presents the best response of player i to strategies ρ_{-i}; and we write $\rho^* = BR(\rho_{-i})$. According to definition 2.1 and definition 2.2 of section 2.1.6.2, we conclude that strategy $Th(\rho^*,\rho_{-i})$ is the pure and unique NE of our coding game. Next, we will examine the benefits that NE can have for CSMA operation and performance.

4.5. CSMA performances at equilibrium

In order to assess the impact of NE on the functioning of the CSMA protocol and on its various performances, we will now proceed with MATLAB simulations of various performances of the CSMA protocol, and make comparisons between equilibrium mode and conventional mode (outside equilibrium). To do this, we will, respectively, address coding cost, output and transmission time. To conclude, we will explain how NE guides the network to better control the energy consumption of its nodes.

4.5.1. *Coding/decoding price (cost)*

Figure 4.5 shows the variation in coding/decoding price (cost) in relation to the redundancy employed by the nodes of the network. It also shows the logical effect of the number of original packets on the assessable value of the price, and which appears in an ascending proportional relationship. The lower cost is achieved when the smaller redundancy is used: $\rho = 2$ packets. At equilibrium, the coder/decoder processes almost 180 mathematical operations to generate the five redundant packets (coder) or recover the nine original packets (decoder). To generate the 14 redundant packets or recover the nine original packets, the coder outside of equilibrium executes 580 mathematical operations. The difference is considerable, and can have an enormous impact on data transmission output. In fact, the difference will be even greater in the real case, with a number on the order of 10,000 mathematical operations to generate or recover a larger number of coded packets. At equilibrium, the estimated cost remains lower while causing the coders/decoders to function more rapidly, which encourages the system to process more data to be transmitted, by increasing the network's capacity and output to the utmost. For the remaining of this chapter, we will consider the number of original packets to be k=9 packets.

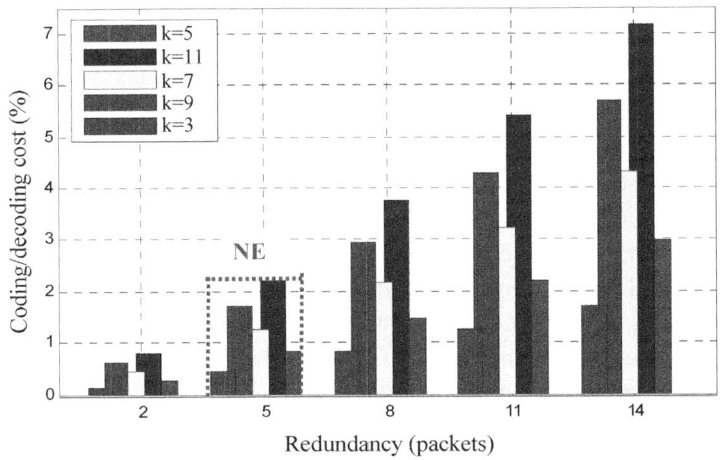

Figure 4.5. *Coding/decoding cost in relation to redundancy and packet size. For a color version of this figure, see www.iste.co.uk/benslama/adhocnetworks.zip*

4.5.2. *Output*

Figure 4.6(b) shows the variation of the probabilities of success and failure of packet transmission in the CSMA network. The two curves vary in inverse proportions and intersect at the average value of the packet in the traffic load. We can see a higher probability of success when traffic is low, which decreases gradually as the traffic load increases. This is logical, given the number of collisions, which seems very small when there is a low traffic load and grows as the traffic load increases.

The variation in transmission output of the CSMA protocol at NE according to the different values of α is shown in Figure 4.6(a), where we can see a considerable improvement compared to the results presented in [TOB 80, KLE 75]. To illustrate the improvement made, in Figure 4.7 we use the value $\alpha = 0.1$ that has been considered in all of our calculations.

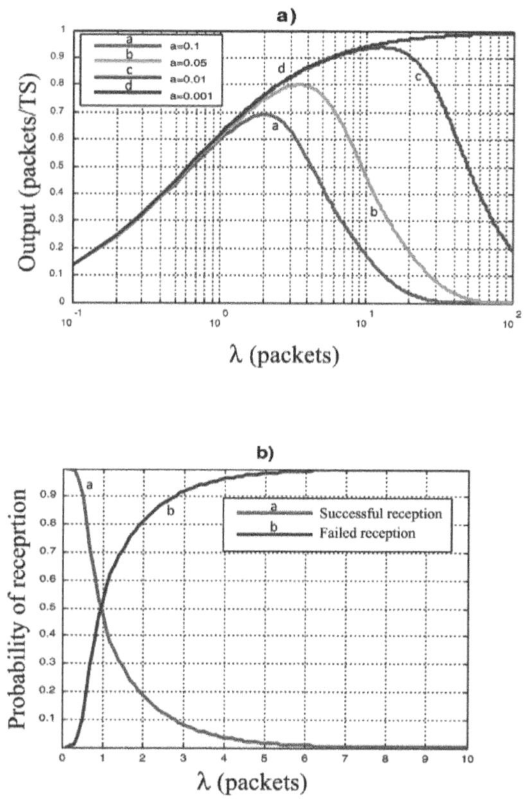

Figure 4.6. *a) CSMA NE output for different values of α. b) Probabilities of transmission success and failure in CSMA NE. For a color version of this figure, see www.iste.co.uk/benslama/adhocnetworks.zip*

In NE, the CSMA protocol functions more rapidly with a higher output than the one seen in conventional mode. Figure 4.7 confirms this, and expresses how our model can be beneficial for CSMA, as it contributes a higher output during a significant margin of the traffic load, so that it is able to reach a maximum peak that is also better than the one attained in conventional mode, with $Th_{max}(NE) = 0.69$

(pack/TS) compared to $Th_{max}(conv) = 0.62$ pack/TS. However, this superiority does not last very long, and the two modes intersect at point (3.2 and 0.62), with conventional CSMA subsequently becoming superior to CSMA at equilibrium. Thus, for relatively low loads, NE brings considerable improvement in the output of the CSMA network.

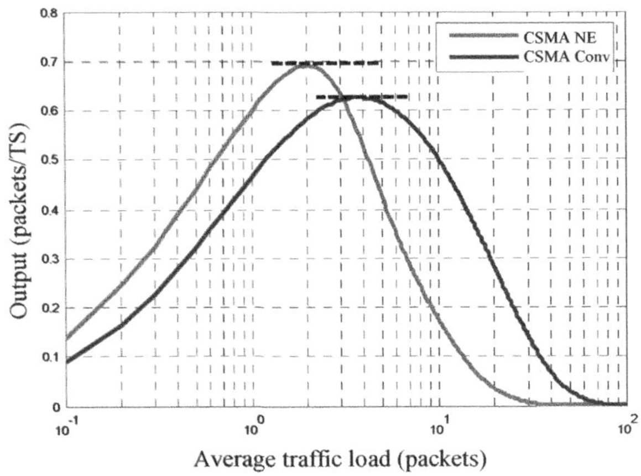

Figure 4.7. *Output of CSMA network at equilibrium and outside equilibrium. For a color version of this figure, see www.iste.co.uk/benslama/adhocnetworks.zip*

4.5.3. *Transmission time*

Total transmission time is defined as being the average time elapsed from the transmission of a packet until its successful arrival at the receiving node; thus in CSMA it includes time spent listening to the channel, transmission time, retransmission time and feedback time.

$$D = D_{listen} + D_{trans} + D_{retrans} + D_{fback} \qquad [4.18]$$

The expression of the total transmission time in a non-persistent CSMA network is given in [KLE 75] by equation:

$$D = (\frac{\lambda}{Th} - 1)(1 + 2\alpha + a + \delta) + 1 + \alpha \qquad [4.19]$$

The time is thus a function of the output, such that $(\lambda/_{Th} - 1)$ represents the average number of retransmissions, $a = \frac{T_{Ack}}{T}$ is the standardized time of the ACK feedback signal and δ is the standardized random period for retransmission. Figure 4.8 shows the transmission times for CSMA at equilibrium (NE) and conventional CSMA. At equilibrium, the CSMA network presents a shorter time compared to the one in conventional mode. This difference in terms of transmission time thus strengthens the CSMA network so that it becomes more efficient by transmitting data in shorter amounts of time. In fact, this benefit provided by NE to the CSMA network is more applicable during the progression phase, given the relatively low traffic loads (up to four packets/TS). Outside of this phase, congestion dominates, and the network (for both modes) ends up in a state of saturation that quickly results in very long transmission times.

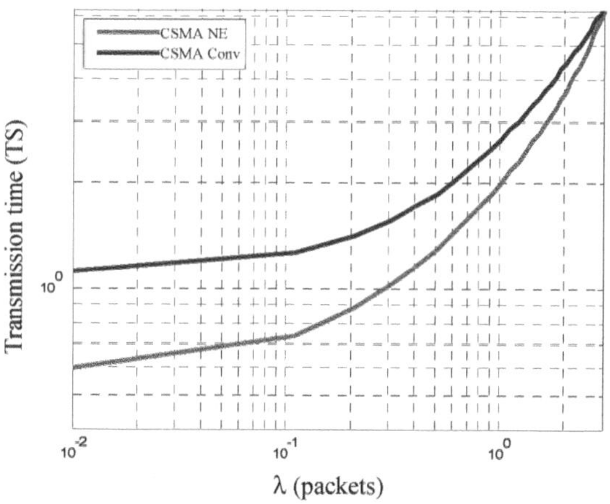

Figure 4.8. *Total transmission time in NE CSMA and conventional CSMA. For a color version of this figure, see www.iste.co.uk/benslama/adhocnetworks.zip*

4.5.4. *Energy optimization at equilibrium*

As we emphasized at the beginning of this chapter, energy control is vital in *ad hoc* networks in order to ensure a long enough duration of service for the completion of all the tasks planned at the outset. We will now explain how our coding game at equilibrium leads the network to a more effective management of its mobile nodes' (players') energy.

Formally, we can see that the relationship between transmission time D and consumed energy E during this transmission is very strong; such that, according to equation [4.18], for a node to transmit its packets successfully, the following time periods must elapse:

$$D = D_{listen} + D_{trans} + D_{retrans} + D_{fback}$$

At the same time, the node consumes energy during each period (time), thus:

$$E = E_{listen} + E_{trans} + E_{retrans} + E_{fback} \quad [4.20]$$

Such that $E_{fback} = E_{ACK}$. We should mention here that the energy consumed during each period includes transmission energy, reception energy, hold energy and the mobility energy of the node in a MANET.

Just before this section, we had begun to discuss the benefits provided by NE to various network performances, particularly with regard to total transmission time, such that at equilibrium the network will experience a higher probability of success with a smaller number of retransmissions, leading to shorter times and lower amounts of energy spent, and thus $D_{retrans} \ll \rightarrow E_{retrans} \ll$. Thus, at equilibrium, the nodes become more efficient in their activities, with a higher output and shorter transmission times resulting in more efficient consumption of available energy. NE, then, leads the network to a more economical consumption of energy.

4.6. Conclusion

The coding game that we have presented in this chapter has enabled us to find an equilibrium strategy that will lead the network to a more stable state of operation. At NE, the CSMA becomes more efficient, with more optimized performances. We have been able to conclude that NE contributes a considerable improvement of various network qualities, and particularly of output and transmission time. Consequently, the CSMA network in equilibrium also optimizes its energy consumption.

A more interesting study is worth developing [GOO 00, ZHE 14, LAI 14, LAN 14, HEI 06, MU 13, XIA 07, LON 11] in this sense, to achieve an optimal optimization of energy.

We have passed over one vital point that would have been a good addition to this chapter; this is the famous problem of network insensitivity to virus attacks [GAO 13, CHO 13, SHE 11, TIA 13].

Conclusion

In this book, we have presented an in-depth study of the random access protocols, slotted ALOHA (SALOHA) and carrier sense multiple access (CSMA), which are used in satellite–terrestrial station links and *ad hoc* networks, respectively. These protocols are characterized by weak and degraded performances due to the collision of packets in transmission. Finally, in this book, we have proposed two complementary solutions: first to increase the probability of successful transmission using a system for the correction and recovery of collided packets, and then to cause the network to maintain its optimum state for a more prolonged period of time. The first solution consists of adding into the transmission system an erasure-coding technique based on Reed–Solomon codes, which is used to correct errors made in the transmission channel, and also to recover on reception all of the original packets sent by the transmitter. Due to this technique, collisions are reduced, relieving the network from finding itself in a process of successive retransmissions, which prolongs transmission times and reduces output. The network thus becomes more efficient with a relatively optimized output. However, the major disadvantage that persists in SALOHA and CSMA protocols lies in their unstable behavior with regard to the arriving traffic load, when their efficiency is restrained to a very narrow interval of the traffic load, and they shift very rapidly into a state of saturation. This is due principally to the random choice of redundancy and also to the egotistical behavior of nodes when each of them seeks to maximize its own profit in a competition to win more free slots in order to transmit the maximum number of packets in the shortest amount of time. For this reason, we have developed all of these conflict situations via a non-cooperative coding game model in order to discuss interactions between nodes and see how to converge toward a common solution able to satisfy all nodes by putting the network in a more stable state of equilibrium. The coding game we have proposed for both SALOHA and CSMA is a non-cooperative repeated game (real case) with imperfect information $J = (\{1,...,M\}, s, u)$. In it, the nodes act as players, and each of them chooses a strategy that maximizes its utility, which is expressed in terms of transmission output. The extraction of the game's utility is the most interesting phase, while the discussion of game equilibrium remains the most difficult and most critical part of game theory. Mathematical analysis of game utility has proven the model's convergence toward a pure and unique solution representing Nash equilibrium (NE) in our coding game.

In order to find and assess the impact of NE on SALOHA and CSMA networks, we have made comparisons between the network in equilibrium mode (NE) and the network outside equilibrium (conventional mode) during simulations of various network performances. We have compared the two modes in the calculation of coding/decoding price (cost); during the assessment of packet loss rates; in the description of output and in the estimation of total transmission time, and we have observed via all these comparisons that the advantage provided by NE is considerable and quite clear in all aspects of the network, particularly in output and transmission time. In fact, the results and simulations have confirmed us that at equilibrium, and with relatively low traffic loads, the network becomes more stable and more efficient, with highly advantageous optimality in terms of packet loss rates, output and transmission time. In addition, the network at equilibrium consumes less energy, which results in a significant level of optimization in the management and control of energy, which also guarantees that *ad hoc* networks will function for longer periods of time, in particular mobile *ad hoc* networks (MANETs), which require additional quantities of energy.

We believe that it is of interest to inform our potential readers that there will be a follow-up to this book, still with the objective of applying game theory to *ad hoc* networks, but this time including multihoming. Our current research in this area has advanced far enough that we can now give the main conclusions we have reached for informational purposes.

We have observed that the inclusion of multihoming in mobile networks, and more specifically in MANETs, makes it possible to increase their robustness and resistance to breakdowns caused by rapid changes in their topologies. Remember that the resources composing the whole chain of communication, from nodes to links in an *ad hoc* network, are limited, which needs the definition of the best way to integrate the multihoming concept by optimizing the information traffic load, especially as high concentrations of this traffic in certain specific nodes cause congestion that can paralyze the entire network.

With the family of proactive routing protocols, a list of the best routes toward all possible destinations is maintained by each node, ensured by the continuous exchange of update messages. In addition, the inclusion of multihoming enables a node to preserve multiple active links at the same time, thus increasing the availability of routes.

Among the direct consequences of this type of configuration are a significant gain in time when a route request is made up for the establishment of a connection. However, a major disadvantage appears when route changes are more frequent than requests, thus causing high levels of partially useless routing traffic, which wastes network capacity and is accompanied by a linear growth of the size of the routing tables depending on the number of nodes and links available.

For the same configuration integrating the concept of multihoming, the protocols of the reactive family, which have the ability to create and maintain routes on request according to need, ensure that no control messages burden the network for unused routes, which makes it so that network resources are not wasted. However, with this routing method, the establishment of a route between the source and the destination requires significant amounts of time, thus causing connection failures.

Given all these disadvantages, we have taken it upon ourselves to propose some suggestions for the optimization of the integration of multihoming into mobile networks in terms of robustness and resistance to the effects of mobility, increased traffic load, route-request response times and the securing of communications.

C.1. Integrating a local route maintenance mechanism into each node

Like that of the *ad hoc* on-demand distance vector (AODV) protocol, which organizes the repair of a route locally when a breakdown is detected by a node, this protocol searches its routing table for an alternate and reliable route, to which it automatically shifts, without transmitting route discovery and maintenance messages. This has the effect of optimizing the routing traffic load and avoiding problems of network congestion.

C.2. Integrating a load-balancing mechanism into autonomous systems

Depending on the study conducted, we have observed the concentration of a heavy load in autonomous systems (ASs), notably those that are near the destination, which can cause congestion problems serious enough to paralyze all traffic. In such a scenario, we propose the use of a load-balancing mechanism based on an indicator of the total traffic load in transit on an AS. This indicator would take into consideration the number of routers making up the AS in question, the routing capacity of each of these and the state of each router. Combining all this information together, the load indicator would be placed at each router with inter-AS links so as to be accessible by all neighboring ASs.

When the load indicator of an AS reaches an average value (defined according to the network mobility model), an analytical process is initiated by this AS in order to detect the one of these neighbors that is the least used and the most likely to extend the destination, in order to shift part of its traffic to the latter, which will take up the relay.

C.3. Reduce response time to route requests

So as not to exceed the average times tolerated by the software, while optimizing the use of network resources, we propose the automatic creation of proximity tables for each AS, refreshing them periodically even in the absence of information packets. This procedure would maintain available inter-AS links and facilitate the procedure of route-finding without wasting network resources, as updates would be executed for the best routes (e.g. Border Gateway Protocol (BGP)).

C.4. Securing the network

In an *ad hoc* network, all nodes are equivalent, and the lack of centralization makes it problematic to access information when an intrusion has been detected.

The first possible solution is to use encryption algorithms, which would require additional processing capacity.

Our suggestion is to integrate into ASs a procedure of identification and authentication of legitimate users via the filtering of medium access control (MAC) addresses, since, statistically speaking, almost all of the nodes in an *ad hoc* network transport their communications at one time or another via an AS; the latter can communicate the identity of a pirate node directly to its neighbors, thus defusing any threat of intrusion.

Finally, to push the limits of mobility even further, beginning with nodes and going up to ASs, the BGP routing protocol has great difficulty in managing proximity tables in the case of a totally mobile network. ASs thus accumulate update messages, leading to congestion problems, which are added to the formation of loops that may disturb the smooth functioning of *ad hoc* routing protocols. Consequently, creating a totally mobile network goes back to eliminating ASs definitively from our topology.

The new routing Steam Control Transmission Protocol (SCTP) appears to be the best solution, using a different method to integrate the concept of multihoming. Based on streams, SCTP enables two entities communicating via an association each to use an Internet Protocol (IP) address list to send multiple waves at once, with the possibility of shifting from a defective link to another link with complete transparency.

Bibliography

[ABR 70] ABRAMSON N., "The ALOHA System: another alternative for computer communications", *AFIPS'70, Proceedings of the November 17-19, 1970, Fall Joint Computer Conference*, pp. 281–285, 1970.

[AGH 06] AGHASSI M., BERTSIAMS D., "Robust game theory", *Mathematical Programming*, Springer, servB, vol. 107, pp. 231–273, 2006.

[AKK 11] AKKARAJITSAKUL K., HOSSAIN E., NIYATO D., *et al.*, "Game theoretic approaches for multiple access in wireless networks: a survey", *IEEE Communications Surveys & Tutorials*, vol. 13, no. 3, pp. 372–395, 2011.

[AKY 02] AKYILDIZ I.F., WEILIAN S., SANKARASUBRAMANIAM Y., *et al.*, "A survey on sensor networks", *IEEE Communications Magazine*, vol. 40, no. 8, pp. 102–116, 2002.

[AKY 05] AKYILDIZ I.F., WANG X., "A survey on wireless meshes networks", *IEEE Communications Magazine*, vol. 43, no. 9, pp. S23–S30, 2005.

[AL 14] AL-DHANHANI A., MIZOUNI R., OTROK H., *et al.*, "A game theoretical model for collaborative groups in social applications", *Expert Systems with Applications*, vol. 41, pp. 5056–5065, 2014.

[ALP 05] ALPCAN T., BASAR T., "Distributed algorithms for Nash equilibria of flow control games", *Advances in Dynamic Games: Applications to Economics, Finance, Optimization, and Stochastic Control*, Annals of the International Society of Dynamic Games, vol. 7, pp. 473–498, 2005.

[ALT 87] ALTMAN E., Evaluation des Files d'attente en station pour le protocole ALOHA, Report no. 698, INRIA, 1987.

[ALT 04] ALTMAN E., EL AZOUZI R., JIMENEZ T., "Slotted ALOHA as a stochastic game with partial information", *Computer Networks*, vol. 45, no. 6, pp. 701–713, 2004.

[ALT 04] ALTMAN E., BARMAN D., EL AZOUZI R., *et al.*, "A game theoretic approach for delay minimization in slotted ALOHA", *Communications IEEE International Conference*, vol. 7, pp. 3999–4003, 2004.

[ANS 14] ANSHELEVICH E., CASKURLU B., KAR K., *et al.*, "Capacity allocation games for network–coded multicast streaming", *IEEE/ACM Transactions on Networking*, vol. 22, no. 2, pp. 595–607, April 2014.

[ANT 10] ANTONIOU J., PAPADOPOULO V., VASSILIOU V., *et al.*, "Cooperative user–network interactions in next generation communication networks", *Computer Networks*, vol. 54, pp. 2239–2255, 2010.

[ANT 14] ANTONOPOULOS A., VERIKOUKIS C., "Multi–player game theoretic MAC strategies for energy efficient data dissemination", *IEEE Transactions on Wireless Communications*, vol. 13, no. 2, p. 592, February 2014.

[ASI 12] ASIOLI S., RAMZAN N., IZQUIERDO E., "A game theoretic approach to minimum-delay scalable video transmission over P2P", *Signal Processing: Image Communication*, vol. 27, pp. 513–521, 2012.

[ATE 08] ATELIN P., *Réseau sans fils 802.11: Technologie, Déploiement, Sécurisation*, 2nd ed., Editions ENI, p. 77, 2008.

[BAC 13] BACCI G., LUISE M., "QoS-aware game-theoretic rate & power control for CDMA wireless communication networks", *Computer Networks*, vol. 57, pp. 1789–1804, 2013.

[BAD 03] BADIS H., EL AGHA K., "QoS for *ad hoc* networking based on multiple–metric: bandwidth and delay", *The 5th IFIP International Workshop On Mobile and Wireless Communications Networks (IFIP MWCN)*, Singapore, October 2003.

[BAE 12] BAE Y.H., CHOI B.D, ALFA A.S., "Achieving maximum throughput in random access protocols with multi-packet reception", *IEEE Transactions on Mobile Computing*, vol. 13, no. 3, pp. 497–511, 2014.

[BAI 12] BAI J., QIU F., XUE Y., End-to-End Fairness over Non-Convex Capacity Region in IEEE 802.11-Based Wireless Networks. *Mobile Adhoc and Sensor Systems (MASS)*, pp. 145–154, 2012.

[BAR 08] BARTH D., BOURNEZ O., BOUSSATON O., *et al.*, "Distributed learning of wardrop equilibria", *Proceedings Unconventional Computing: 7th International Conference*, Series: Lecture Notes in Computer Sciences, Springer, pp. 19–32, 2008.

[BAR 14] BARTAL P., NAGY I., "Game theoretic approach for achieving optimum overall efficiency in DC/DC converters", *IEEE Transactions on Industrial Electronics*, vol. 61, no. 7, p. 3302, July 2014.

[BAS 94] BASU K., "The traveler's dilemma: paradoxes of rationality in game theory", *AEA papers and proceedings Rationality and Well-Being*, vol. 84, no. 2, pp. 391–395, 1994.

[BEN 07] BENSAOU B., FANG Z., "A fair MAC protocol for IEEE 802.11-based *ad hoc* networks: design and implementation, *IEEE Transactionson Wireless Communications*, vol. 6, no. 8, p. 2934, August 2007.

[BER 87] BERTSEKAS D., GALLAGER R., *Data Networks*, Prentice Hall, Englewood Cliffs, NJ, 1987.

[BIN 85] BINMORE K., SHAKED A., SUTTON J., "Association testing non cooperative bargaining theory: a preliminary study source", *The American Economic Review*, vol. 75, no. 5, pp. 1178–1180, December1985.

[BIN 05] BINI D.A., MEINI B., LATOUCHE G., *Numerical Methods for Structured Markov Chains*, Oxford Science publishing 2005.

[BIR 14] BIRJE M.N., SUNILKUMAR S.M., DAS S.K., "Reliable resources brokering scheme in wireless grids based on non-cooperative bargaining game", *Journal of Network and Computer Applications*, vol. 39, pp. 266–279, 2014.

[BOR 21] BOREL E., "La Théorie du jeu et les équations intégrales à noyau symétrique gauche", *Comptes Rendus Académie des Sciences*, vol. 173, pp. 1304–1308, 1921.

[BRO 08] BROWN K.L., SHOHAM Y., *Essentials of Game Theory*, Morgan and Claypool Publishers, 2008.

[CAG 05] CAGALJ M., GANERIWAL S., AAD I., *et al.*, "On selfish behavior in CSMA/CA networks", *INFOCOM, 24th annual joint conference of the IEEE Computer and Communications Societies*, vol. 4, pp. 2513–2524, 2005.

[CAM 05] CAMERER C.F., *Behavioral Game Theory, Experiments in Strategic Interaction*, Princeton Press, 2005.

[CAN 00] CANO J.C., MANZONI P., "A performance comparison of energy consumption for mobile *ad hoc* network routing protocols", *Modeling, Analysis and Simulation of Computer and Telecommunication Systems Proceedings 8th International Symposium*, pp. 57–64, 2000.

[CAR 99] CARSON S., MACKER J., Mobile *ad hoc* networking (manet):routing protocol performance issues and evaluation Considerations. RFC n2501, IETF, 1999.

[CAR 03] CARTIGNY J., INGELREST F., SIMPLOT D., "RNG relay subset flooding protocols in mobile *ad hoc* networks", *International Journal of Foundations of Computer Science*, vol. 14, no. 2, pp. 253–265, 2003.

[CHA 04] CHAN D.S., BERGER T., BRIDGELALL R., "Energy efficiency of CSMA protocols for wireless packet switched networks", *Proceedings of IEEE Wireless Communications and Networking Conference (WCNC)*, vol. 1, pp. 447–452, 2004.

[CHA 10] CHARILAS D.E., PANAGOPOULOS A.D., "A survey on game theory applications in wireless networks", *Computer Networks*, vol. 54, pp. 3421–3430, 2010.

[CHA 14a] CHAI B., YANG Z., "Impacts of unreliable communication and modified Regret matching based anti-jamming approach in smart microgrid", *Ad Hoc Networks*, vol. 22, pp. 69–82, 2014.

[CHA 14b] CHAVALI P., MAHORAIS A., "Hierarchical particle filtering for multi-modal data fusion with application to multiple-target tracking", *Signal Processing*, vol. 97, pp. 207–220, 2014.

[CHE 98] CHEN T.W., GERLA M., "Global state routing: a new routing scheme for ad-hoc wireless networks", *Communications, ICC 98 Conference record 1998, IEEE International conference on Communications*, vol. 1, pp. 171–175, 1998.

[CHE 01] CHEN B., JAMIESON K., BALAKRISHNAN H., *et al.*, "Span: an energy efficient coordination algorithm for topology maintenance in *ad hoc* wireless networks", *Wireless Networks*, vol. 8, pp. 481–494, 2002.

[CHE 03] CHEON T., "Altruistic duality in evolutionary game theory", *Physics Letters A*, vol. 318, pp. 327–332, 2003.

[CHE 11] CHEN Z., YIHUI Q., JINGJING L., *et al.*, "Incentive mechanism for selfish nodes in wireless sensor networks based on evolutionary game", *Computers and Mathematics with Applications*, vol. 62, pp. 3378–3388, 2011.

[CHE 13] CHEUNG M.H., WONG V.W.S., "Interference pricing for SINR based random access game", *IEEE Transactions on Wireless Communications*, vol. 12, no. 5, p. 2292, May 2013.

[CHI 05] CHINTALAPUDI K.K., DHARIWAL A., GOVINDAN R., *et al.*, *Ad Hoc* Localization Using Ranging and Sectoring, *INFOCOM 2004, Twenty Third annual Joint Conference of the IEEE Computer and communication Societies*, vol. 4, pp. 2662–2672, 2004.

[CHO 13] CHORPPATH A.K., ALPCAN T., "Trading privacy with incentives in mobile commerce: a game theoretic approach", *Pervasive and Mobile Computing*, vol. 9, pp. 598–612, 2013.

[CHU 14] CHUNGETAL S.H., "Dynamic sustainability games for renewable resources: a computational approach", *IEEE Transactions on Computers*, vol. 63, no. 1, p. 155, January 2014.

[COR 99] CORSEN S., MACKER J., Mobile *Ad Hoc* Networking (MANET): Routing Protocol Performance Issues and Evaluation Consideration, RFC 2501, 1999.

[DAN 99] DANDACHE A., VALLINO T., MONTEIRO F., *et al.*, "Code Reed-Solomon avec effacements : simulation et conception sur réseaux de circuits programmables (FPGA)", *Revue Traitement du Signal*, vol. 16, no. 4, pp. 332–341, 1999.

[DER 12] DERAKHSHANI M., NGOC T.L., "Intelligent CSMA-based opportunistic spectrum access: competition and cooperation", *IEEE Global communication Conference* (*Globecom)* pp. 1386–1390, 2012

[DEX 14] DEXTREIT C., KOLMANOVSKY I.V., "Game theory controller for hybrid electric vehicles", *IEEE Transactions on Control SystemsTechnology*, vol. 22, no. 2, p. 652, March 2014.

[DIX 10] DIXIT A.K., NALEBUFF B.J., *The Art of Strategy: A Game Theorist's Guide to Success in Business and Life*, 2010.

[DRE 61] DRESHER M., *Games of Strategy, Theory and Applications*, Prentice Hall Inc., 1961.

[DOR 14] DORRONSORA B., RUIZ P., DANOY G., *et al.*, *Evolutionary Algorithms for Mobile Ad Hoc Networks*, John Wiley and Sons, 2014.

[DOU 02] DOUCEUR J.R., *The Sybil Attack*, Lecture Notes In Computer Science, Springer, vol. 2429, pp. 251–260, 2002.

[DUA 14] DUAN J., DEYING LI., WENPING C., *et al.*, "3D geometric routing without loops and dead ends in wireless sensor networks", *Ad Hoc Networks*, vol. 13, pp. 312–320, 2014.

[EBE 13] EBER N., *Les Théories des Jeux*, Dunod, 2013.

[EL 03] EL-AZOUZI R., ALTMAN E., WYNTER L., "Telecommunications network equilibrium with price and quality-of-service characteristics", *Teletraffic Science and Engineering*, vol. 5, pp. 369–378, 2003.

[EL 13] EL-AZOUZI R., DE PELEGRINI F., SIDI H., *et al.*, "Evolutionary forwarding games in delay tolerant networks: equilibria, mechanism design and stochastic approximation", *Computer Networks*, vol. 57, pp. 1003–1018, 2013.

[ELI 11] ELIAS J., MARTIGNON F., CAPONE A., *et al.*, "Non-cooperative spectrum access in cognitive radio networks: a game theoretical model", *Computer Networks*, vol. 55, pp. 3832–3846, 2011.

[FOR 99] FORGO F., SZEP J., SZIDAROVSZKHY F., *Introduction to the Theory of Games: Concepts, Methods, Applications*, Springer, 1999.

[FEE 01] FEENEY L. M., NILSSON M., "Investigating the energy consumption of a wireless network interface in an *ad hoc* networking environment", *Proceedings of IEEE INFOCOM*, vol. 3, pp. 1548–1557, 2001.

[FUD 91] FUDENBERG D., TIROLE J., *Game Theory*, The MIT Press, Cambridge, MA, 1991.

[FUD 98] FUDENBERG, LEVINE, *The Theory of Learning in Games*, The MIT Press, Cambridge, MA, 1998.

[GAD 04] GADALLAH Y., KUNZ T., "PIES: protocol independent energy saving algorithm", *International Conference on Parallel ProcessingWorkshops (ICPPW'04)*, pp. 4–12, 2004.

[GAJ 14] GAJIĆ V., JIANWEI H., RIMOLDI B., "Competition of wireless providers for atomic users", *IEEE/ACM Transactions on Networking*, vol. 22, no. 2, p. 512, April 2014.

[GÁL 12] GÁLLEGO J.R., CANALES M., ORTIN J., "Distributed resource allocation in cognitive radio networks with a game learning approach to improve aggregate system capacity", *Ad Hoc Networks*, vol. 10, pp. 1076–1089, 2012.

[GAO 13] GAO P., MENG D.X., CHENG N., *et al.*, "Non-cooperative power control game for adaptive modulation and coding", The Journal of China Universities of Posts and Telecommunications, vol. 17, no. 3, pp. 31–37, June 2010.

[GAS 13] GASIOR D., DERWAL M., "Pareto-optimal Nash equilibrium in capacity allocation game for self-managed networks", *Computer Networks*, vol. 57, pp. 2817–2832, 2013.

[GHA 10] GHASEMIAND A., FAEZ K., "A non-cooperative game approach for power-aware MAC in *ad hoc* wireless networks", *Computer Communications*, vol. 33, pp. 1440–1451, 2010.

[GHA 13a] GHARESIFARD B., CORTÉS J., "Distributed convergence to Nash equilibria in two-network zero-sum games", *Automatica*, vol. 49, pp. 1683–1692, 2013.

[GHA 13b] GHAZVINI M., MOVAHEDINIA N., JAMSHIDI K., *et al.*, "Game theory applications in CSMA methods", *IEEE Communications Surveys & Tutorial*s, vol. 15, no. 3, pp. 1062–1087, 2013.

[GIB 92] GIBBONS R., *Game Theory for Applied Economists*, 1992.

[GOO 00] GOODMAN D., MANDAYAM N., "Power control for wireless data", *IEEE Personal Communications*, vol. 7, no. 2, pp. 48–54, April 2000.

[GUE 10] GUERRIEN B., *La théorie des jeux*, Eyrolles 2010.

[GUO 11] GUO L., WU J., HOU W., *et al.*, "Hybrid protection algorithms based on game theory in multi-domain optical networks", *Optical Fiber Technology*, vol. 17, pp. 523–535, 2011.

[GUP 97] GUPTA P., KUMAR P.R., "A system and traffic dependent adaptive routing algorithm for *ad hoc* networks", *Decision and control 1997, Proceedings of the 36th IEEE Conference on Decision and control*, vol. 3, pp. 2375–2380, 1997.

[GYA 11] GYARMATIAND L., TRINH T.A., "Cooperative strategies of wireless access technologies: a game-theoretic analysis", *Pervasive and Mobile Computing*, vol. 7, pp. 545–568, 2011.

[HAC 03] HAC A., *Mobile Telecommunications Protocols for Data Networks*, John Wiley and Sons, 2003.

[HAN 04] HANZO L., LIEW T.H., YEAP B.L., *Turbo Coding, Turbo Equalization and Space Time Coding for Transmission Over Fading Channels*, John Wiley and Sons, New York, 2004.

[HAN 08] HAN Z., LIU K.J.R., *Resource Allocation for Wireless Networks:Basics, Techniques, and Applications*, Cambridge University Press, 2008.

[HAN 12] HAN Z., NIYATO D., BASAR T., *et al.*, *Game Theory in Wireless and Communication Networks: Theory, Models, and Applications*, Cambridge University Press, Cambridge, MA, 2012.

[HAR 99] HARMER G.P., ABBOTT D., "Losing strategies can win by Parrondo's paradox", *Nature*, vol. 402, 23/30 December 1999. Available at: www.nature.com.

[HAR 04] HARA B.O., *IEEE 802.11 Handbook: A Designer's Companion*, Institute of Electrical Electronics Engineers (IEEE), 2004.

[HAR 10] HARRINGTON J.E., GAMES J.R., *Strategies, and Decision Making*, Worth Publishing, 2010.

[HEC 05] HECKER A., Contrôle d'accès et gestion des réseaux 4G hétérogènes, Ecole Nationale Supérieure des Télécommunications, Paris, 2005.

[HEI 06] HEIKKINEN T., "A potential game approach to distributed power control and scheduling", *Computer Networks*, vol. 50, pp. 2295–2311, 2006.

[HOU 14] HOU R., WANG C., ZHU Q., *et al.*, "Interference-aware QoS multicast routing for smart grid", *Ad Hoc Networks*, vol. 22, pp. 13–26, 2014.

[IGA 11] IGARTUA M.A., DE LA CRUZ L.J., FRIAS V.C., *et al.*, "A game-theoretic multipath routing for video-streaming services over mobile *ad hoc* networks", *Computer Networks*, vol. 55, pp. 2985–3000, 2011.

[ILT 06] ILTIS R.A., SEUNG J.K., HOANG D.A., "Non cooperative iterative MMSE beam forming algorithms for *ad hoc* networks", *IEEE Transactions on Communications*, vol. 54, no. 4, pp. 748–759, 2006.

[JAC 14] JACKSON C.A., RUSSELL H.B, "A channel-access protocol to utilize multiple heterogeneous channels for *ad hoc* networks", *Ad Hoc Networks*, vol. 13, pp. 245–255, 2014.

[JIA 09] JIANG L., WALRAND J., "Approaching throughput-optimality in a distributed CSMA algorithm: collisions and stability", *ACM Mobihoc'09 S3 Workshop*, 2009.

[JIA 14] JIANG C., CHEN Y., LIU K.J.R., "On the equivalence of evolutionary stable strategies", *IEEE Communications Letters*, vol. 18, no. 6, June 2014.

[JON 01] JONES C.E., SIVALINGAM K.M., AGRAWAL P., *et al.*, "A survey of energy efficient network protocols for wireless networks", *Wireless Networks*, vol. 7, no. 4, pp. 343–358, 2001.

[JOR 09] JORSWIECK E.A., LARSSON E.G., LUISE M., *et al.*, "Game theory in signal processing and communication", *IEEE Signal Processing Magazine*, vol. 17, September 2009.

[JUB 87] JUBIN J., TORNOW J.D., "The darpa packet radio network protocols", *Proceedings of the IEEE*, pp. 21–32, 1987.

[JUL 12] JULMI C., *Introduction to Game Theory*, Bookboon.com, 2012.

[JUR 07] JURDAK R., *Wireless Ad Hoc and Sensor Networks: A Cross-Layer Design Perspective*, Springer, 2007.

[KAC 12] KACKER F., BOUIBED K., *La Théorie des Jeux: Cours, Exercices et Sujets d'Examens corrigés*, Pages Bleues Internationales, 2012.

[KAL 10] KALAM S., MAHBUB G., SENEVIRATNE L., "A game-theoretic approach to non-cooperative target assignment", *Robotics and Autonomous Systems*, vol. 58, pp. 955–962, 2010.

[KAR 90] KARN P., "MACA a new channel access method for packet radio", *ARRL/CRRL Amateur Radio 9th Computer Networking Conference*, London, Canada, pp. 134–140, 1990.

[KAR 14] KARAHAN A., ERTURK I., ATMACA S., *et al.*, "Effects of transmit-based and receive-based slot allocation strategies on energy efficiency in WSN MACs", *Ad Hoc Networks*, vol. 13, pp. 404–413, 2014.

[KIN 05] KING-SUN C., YEUNG L.K., SHAO W., "Contention-based MAC protocols with erasure coding for wireless data networks", *Ad hoc Networks*, vol. 3, no. 4, pp. 495–506, 2005.

[KLE 75] KLEINROCK L., TOBAGI F.A., "Carrier sense multiple access for packet switched radio channel", *IEEE Transactions on Communications*, vol. 23, no. 12, pp. 1400–1416, 1975.

[KOM 08] KOMATHY K., NARAYANASAMY P., "Trust-based evolutionarygame model assisting AODV routing against selfishness", *Journal ofNetwork and Computer Applications*, vol. 31, pp. 446–471, 2008.

[KON 05] KONORSKY J., "Solvability of a Markovian model of an IEEE802.11 LAN under a backoff attack", *Proceedings of IEEE MASCOTS*, Atlanta, pp. 491–498, 2005.

[KON 06] KONORSKY J., "A game-theoretic study of CSMA/CA under a backoff attack", *IEEE/ACM Transactions on Networking*, vol. 14, no. 6, 2006.

[KON 10] KONG Z., KWOK YU-K., "Efficient wireless packet scheduling in a non-cooperative environment: game theoretic analysis and algorithms", *Journal of Parallel and Distributed Computing*, vol. 70, pp. 790–799, 2010.

[KOS 05] KOSKIE S., GAJIC Z., "A Nash game algorithm for SIR-based power control in 3G wireless CDMA networks", *IEEE/ACM Transactionson Networks*, vol. 13, no. 5, pp. 1017–1026, 2005.

[KOU 14] KOUTSOPOULOS I., TASSIULAS L., GKATZIKIS L., "Client–server games and their equilibria in peer-to-peer networks" *Computer Networks*, vol. 67, pp. 201–218, 2014.

[KRA 98] KRAVETS R., KRISHNAN P., "Power management techniques for mobile communications", *Proceedings of the ACM Mobile Computing and Networking Conference*, Dallas, Texas, pp. 157–168, 1998.

[LAI 14] LAI H.-Q., CHEN Y., RAY LIU K.J., "Energy efficient cooperative communications using coalition formation games", *Computer Networks*, vol. 58, pp. 228–238, 2014.

[LAN 14] LANGAR R., SECCI S., BOUTABA R., *et al.*, "An operations research game approach for resource and power allocation in cooperative femtocell networks", *IEEE Transactions on Mobile Computing*, vol. PP, no. 99, 2014.

[LEE 00] LEE C., *Error Control Bloc Codes*, Artech House Telecommunications, 2000

[LEE 14] LEE K.H., CHO S., KIM J.H, "Performance evaluation of network coding in IEEE 802.11 wireless *ad hoc* networks", *Ad Hoc Networks*, vol. 16, pp. 131–141, 2014.

[LES 12] LESHEM A., ZEHAVI E., YAFFE Y., "Multichannel opportunistic carrier sensing for stable channel access control in cognitive radio systems", *IEEE Journal on Selected Areas in Communications*, vol. 30, no. 1, p. 82, January 2012.

[LET 97] LETTIERI P., FRAGOULI C., SRIVASTAVA M.B., "Low power error control for wireless links", MobiCOM', *Proceedings of the 3 rd Annual ACM/IEEE International Conference on Mobile Computing and Networking*, Budapest, Hungary, pp. 139–150, 1997.

[LI 01] LI L., HALPEM J.Y., "Minimum-energy mobile wireless networks", *IEEE International Conference on Communications*, 2001.

[LI 14] LI F., WANG L., LU W., "Potential bargaining for resource allocation in cognitive relay transmission", *Journal of Network and Computer Applications*, vol. 37, pp. 253–258, 2014.

[LIN 04] LIN S., COSTELLO D.J. Jr., *Error Control Coding*, Prentice Hall, Englewood Cliffs, 2004.

[LIN 14] LIN W., "Distributed UAV formation control using differentialgame approach", *Aerospace Science and Technology*, vol. 35, pp. 54–62, 2014.

[LIU 11] LIU W.-Y., YUE K., WU T.Y., *et al.*, "An approach for multi-objective categorization based on the game theory and Markov process", *Applied Soft Computing*, vol. 11, pp. 4087–4096, 2011.

[LON 11] LONG C., GUAN X., "Joint random access and power control game in *ad hoc* networks with non cooperative users", *Ad Hoc Networks* vol. 9, pp. 142–151, 2011.

[LOP 07] LOPEZ L., FERNANDEZ A., CHOLVI V., "A game theoretic comparison of TCP and digital fountain based protocols", *Computer Networks*, vol. 51, pp. 3413–3426, 2007.

[LUC 59] LUCE R.D., RAIFFA H., *Games and Decisions: Introduction and Critical Survey*, Dover Publications, 1959.

[MA 06] MA R.T.B., MISRA V., RUBENSTEIN D., "Modeling and analysis of generalized slotted ALOHA MAC protocols in cooperative, competitive and adversarial environments", *Distributed Computing Systems, 26th International Conference*, pp. 62–66, 2006.

[MAC 01] MACKENZIE A., WICKER S.B., "Selfish users in ALOHA: a game theoretic approach", *Proceedings of Fall IEEE Vehicular Technologyconference*, vol. 3, pp. 1354–1357, 2001.

[MAC 03] MACKENZIE A., WICKER S., "Stability of multipacket slotted ALOHA with selfish users and perfect information", *Proceedings of IEEE INFOCOM*, vol. 3, pp. 1583–1590, 2003.

[MAC 08] MACHADO R., TEKINAY S., "A survey of game-theoreticapproaches in wireless sensor networks", *Computer Networks*, vol. 52, pp. 3047–3061, 2008.

[MAK 07] MAK P.I., PAN U.S, MARTIN R.P., *Analog Baseband Architectures and Circuits for Multistandard and Low Voltage Wireless Transceivers*, Springer, 2007.

[MAR 00] MARTI S. GIULI T.J., LAI K., *et al.*, "Mitigating routing misbehavior in mobile *ad hoc* networks", MOBICOM '00, *Proceedings of the 6th Annual International Conference on mobile computing and Networking*, pp. 255–265, 2000.

[MAS 09] MASKERY M., KRISHNAMURTHY V., QING Z., "Decentralized dynamic spectrum access for cognitive radios: cooperative design of a non-cooperative game", *IEEE Transactions on Communications*, vol. 57, no. 2, pp. 459, February 2009.

[MEJ 11] MEJIA M., PENA N., MUNOZ J.L., *et al.*, "A game theoretic trust model for on-line distributed evolution of cooperation in MANETs", *Journal of Network and Computer Applications*, vol. 34, pp. 39–51, 2011.

[MEN 05] MENASCH D.S., FIGUEIREDO D.R., DE SOUZA E.S., "An evolutionary game-theoretic approach to congestion control", *Performance Evaluation*, vol. 62, pp. 295–312, 2005.

[MES 11] MESSAI A., SADOUNI S., BENSLAMA M., *et al.*, "The integration of the multihoming concept in *Ad hoc* Manet mobile networks", *PIERS Proceeding*, *Marrakech*, Morocco, pp. 1229–1234, March 20–23, 2011.

[MOC 12] MOCHAOURAB R., JORSWIECK E.A., "Robust beamforming in interference channels with imperfect transmitter channel information", *Signal Processing*, vol. 92, pp. 2509–2518, 2012.

[MOD 96] MODERER D., SHAPLEY L.S., "Potential games", *Games and Economic Behavior*, vol. 14, no. 1, pp. 124–143, 1996.

[MOK 14] MOKDAD L., BEN OTHMAN J., YAHYA B., *et al.*, "Performance evaluation tools for QoS MAC protocol for wireless sensor networks", *Ad Hoc Networks*, vol. 12, pp. 86–99, 2014.

[MU 13] MU S., ZHU QI, "Power distribution algorithm based on game theory in the femtocell system", *The Journal of China Universities of Posts and Telecommunications*, vol. 20, no. 2, pp. 42–47, April 2013.

[MÜH 02] MÜHLETHALER P., *802.11 et les réseaux sans fils*, Eyrolles, Paris, p. 281, 2002.

[MUH 04] MUHLETHAL P., SALAUN M., QAYYUM A., *et al.*, Comparison between ALOHA and CSMA in Multiple hop *ad hoc* Networks, Report, no. 5129, INRIA, 2004.

[MUK 14] MUKHERJEE A., FAKOORIAN S.A.A., JING H., *et al.*, "Principles of physical layer security in multiuser wireless networks: a survey", *IEEE Communications Surveys &Tutorials*, vol. 16, no. 3, pp. 1550–1573, 2014.

[MYE 91] MYERSON R., *Game Theory: Analysis of Conflict*, Harvard University Press, Cambridge, MA, 1991.

[NAS 51] NASH J., "Non cooperative games", *Annals of Mathematics*, vol. 54, pp. 289–295, 1951.

[NAS 53] NASH J., "Two person cooperatives games", *Econometrica*, vol. 21, pp. 128–140, 1953.

[NIS 07] NISAN N., ROUGHGARDEN T., TARDOS E., *et al.*, *Algorithmic Game Theory*, Cambridge University Press, 2007.

[NIY 07] NIYATO D., HOSSAIN E., "QoS-aware bandwidth allocation and admission control in IEEE 802.16 broadband wireless access networks: anon-cooperative game theoretic approach", *Computer Networks*, vol. 51, pp. 3305–3321, 2007.

[NOV 14] NOVALES R.L., MITTAL N., SARAC K., "SKAIT: a parameterized key assignment scheme for confidential communication in resource constrained *ad hoc* wireless networks", *Ad Hoc Networks*, vol. 20, pp. 163–181, 2014.

[ODE 10] ODERANTIAND F.O., DE WILDE P., "Dynamics of business games with management of fuzzy rules for decision making", *InternationalJournal of Production Economics*, vol. 128, pp. 96–109, 2010.

[OSB 94] OSBORNE, RUBINSTEIN, *A Course in Game Theory*, The MIT Press, Cambridge, MA, 1994.

[OSB 00] OSBORNE M.J., *An Introduction to Game Theory*, Oxford University Press, 2000.

[PAL 10] PALOMAR D.P., ELDAR Y.C., *Convex Optimization in Signal Processing and Communications*, Cambridge University Press, 2010.

[PAR 09] PARK K.J., HOU J.C., BASAR T., *et al.*, "Non cooperative carrier sense game in wireless networks", *IEEE Transactions on Wireless Communications*, vol. 8, no. 10, pp. 5280–5289, October 2009.

[PAT 03] PATEUX S., LE GUELVOUIT G., "Practical watermarking scheme based on wide spread spectrum and game theory", *Signal Processing: Image Communication*, vol. 18, pp. 283–296, 2003.

[PAV 14] PAVKOVIC B., DUDA A., HWANG W.J., *et al.*, "Efficient topology construction for RPL over IEEE 802.15.4 in wireless sensor networks", *Ad Hoc Networks*, vol. 15, pp. 25–38, 2014.

[PEN 10] PENG G., MENG D.X., CHENG N., *et al.*, "Non-cooperative power control game for adaptive modulation and coding", *The Journal of China Universities of Posts and Telecommunications*, vol. 17, no. 3, pp. 31–37, June 2010.

[PER 03] PERKINS C., BELDING ROYER E., DAS S., *Ad Hoc* on-demand distance vector (AODV) routing, RFC 3561, IETF, 2003.

[PER 12] PERLAZA S.M., TEMBINE H., LASAULCE S., *et al.*, "Quality-of-service provisioning in decentralized networks: a satisfaction equilibrium approach", *IEEE Journal of Selected Topics in Signal Processing*, vol. 6, no. 2, pp. 104–116, April2012.

[PIL 14] PILLONI V. NAVARATNAM P., VURAL S., *et al.*, "TAN: a distributed algorithm for dynamic task assignment in WSNs", *IEEE Sensors Journal*, vol. 14, no. 4, pp. 1266, April 2014.

[QUE 04] QUESETH O., "Cooperative and selfish behaviour in unlicensed spectrum using the CSMA/CA protocol", *Proc Nordic Radio Symp*, Oulu, Finland 2004.

[QUE 05] QUESETH O., "The effect of selfish behavior in mobile networks using CSMA/CA", *IEEE Vehicular Technology Conference*, VTC 2005-Spring, vol. 4, pp. 2157–2161, 2005.

[PRA 03] PRASAD R., RUGGIERI M., *Technology Trends in Wireless Communications*, Artech House, 2003.

[RAM 14] RAMASWAMY V., CHOUDHURY D., SHAKKOTAI S., "Which protocol? Mutual interaction of heterogeneous congestion controllers", *IEEE/ACM Transactions on Networking*, vol. 22, no. 2, p. 457, April 2014.

[RAS 01] RASMUSSEN E., *Jeux et informations: Introduction à la théorie des jeux*, De Boeck, 2003.

[RAZ 14] RAZZAQUE M.A., AHMED M.H.U., HONG C.S., *et al.*, "QoS-aware distributed adaptive cooperative routing in wireless sensor networks", *Ad Hoc Networks*, vol. 19, pp. 28–42, 2014.

[REN 11] RENNAAND P., ARGONETO P., "Capacity sharing in a network of independent factories: a cooperative game theory approach", *Robotics and Computer-Integrated Manufacturing*, vol. 27, pp. 405–417, 2011.

[ROB 72] ROBERTS L.G., ALOHA Packet System with and without Slots and Capture, ASS note 8, Stanford, CA: Stanford research institute, *Advanced research projects Agency*, Network information Center, 1972.

[ROH 14] ROH H.T., LEE J.W., "Distributed node placement algorithm utilizing controllable mobility in mobile *ad hoc* networks", *Ad Hoc Networks*, vol. 15, pp. 67–77, 2014.

[ROS 73] ROSENTHAL R.W., "A class of games possessing pure-strategy Nash Equilibria", *International journal of Game Theory*, pp. 65–67, 1973.

[RUN 10] RUNYAN P., HEPING W., ZHENGPING W., *et al.*, "Decision-making of aircraft optimum configuration utilizing multidimensional game theory", *Chinese Journal of Aeronautics*, vol. 23, pp. 194–197, 2010.

[SAB 11] SABIR E., EL AZOUZI R., HAYEl Y., "Hierarchy sustains partial cooperation and induces a Braess-like paradox in slotted ALOHA based networks", *Computer Communications Journal*, vol. 35, no. 3, pp. 273–286, 2012.

[SAH 14] SAHIN D., GUNGOR V.C., KOCAK T., *et al.*, "Quality-of-service differentiation in single-path and multi-path routing for wireless sensor network-based smart grid applications", *Ad Hoc Networks*, vol. 22, pp. 43–60, 2014.

[SCH 90] SCHACHAM N., MCKENNY P., "Packet recovery in high-speed networks using coding and buffer management", *Infocom*, pp. 124–131, 1990.

[SCH 01] SCHURGERS C., TSIATSIS V., SRIVASTAVA M.B., "STEM: topology management for energy efficient sensor networks", *IEEE Aerospace Conference proceedings*, vol. 3, pp. 1099–1108, 2002.

[SEN 14] SENTURK I.F., AKKAYAKA K., YILMAZ S., Relay placement for restoring connectivity in partitioned wireless sensor networks under limited information", *Ad Hoc Networks*, vol. 13, Part B, pp. 487–503, 2014.

[SHA 13] SHARMA R.S., BHATTACHARYA S., "Knowledge dilemmas within organizations: resolutions from game theory", *Knowledge-Based Systems*, vol. 45, pp. 100–113, 2013.

[SHA 14] SHAH G.A., AKAN O.B., "Performance analysis of CSMA-based opportunistic medium access protocol in cognitive radio sensor networks", *Ad Hoc Networks*, vol. 15, pp. 4–13, 2014.

[SHE 11] SHEN S., LI Y., XU H., *et al.*, "Signaling game based strategy of intrusion detection in wireless sensor networks", *Computers and Mathematics with Applications*, vol. 62, pp. 2404–2416, 2011.

[SHE 12] SHEN S., HAN R., GUO L., *et al.*, "Survivability evaluation towards attacked WSNs based on stochastic game and continuous-time Markov chain", *Applied Soft Computing*, vol. 12, pp. 1467–1476, 2012.

[SIN 98] SINGH S., RAGHAVENDRA C.S., "Power aware multi-access protocolwith signaling for *ad hoc* networks", *ACM Computer Communication Review*, vol. 28, no. 3, pp. 5–26, 1998.

[SIV 00] SIVALINGAM K.M., CHEN J.C., AGRAWAL P., *et al.*, "Design and analysis of low power access protocols for wireless and mobile ATM networks", *Wireless Networks*, vol. 6, no. 1, pp. 73–87, 2000.

[SNI 85] SNIDAL D., *The Game Theory of International Politics: Source:World Politics*, Cambridge University Press Stable, vol. 38, no. 1, pp. 25–57, October 1985.

[SON 12] SONG L., HAN Z., ZHANG Z., *et al.*, "Non-cooperative feedback-rate control game for channel state information in wireless networks", *IEEE Journal on Selected Areas in Communications*, vol. 30, no. 1, pp. 188, January 2012.

[SOR 14] SORYAL J., SAADAWI T., "IEEE 802.11 DoS attack detection and mitigation utilizing cross layer design", *Ad Hoc Networks*, vol. 14, pp. 71–83, 2014.

[STE 97] STEMM M., KATZ R.H., "Measuring and reducing energy consumption of network interfaces in hand held devices", *IEICE Transactions on Communications*, vol. E80-B(8), pp. 1125–1131, 1997.

[SU 07] SU B.B., CHANG H., CHEN Y.Z., *et al.*, "A game theory model of urban public traffic networks", *Physica A*, vol. 379, pp. 291–297, 2007.

[SU 14] SU W., HUANG A.Q., "A game theoretic framework for a next generation retail electricity market with high penetration of distributed residential electricity suppliers", *Applied Energy*, vol. 119, pp. 341–350, 2014.

[SWA 14] SWAIN P., CHAKRABORTY S., NANDI S., *et al.*, "Performance modeling and evaluation of IEEE 802.11 IBSS power save mode", *Ad Hoc Networks*, vol. 13, pp. 336–350, 2014.

[SZA 07] SZABO G., FATH G., "Evolutionary games on graphs", *Physics Reports*, vol. 446, pp. 97–216, 2007.

[SZO 14] SZOTT K.K., "A comprehensive analysis of IEEE 802.11 DCF heterogeneous traffic sources", *Ad Hoc Networks*, vol. 16, pp. 165–181, 2014.

[TAV 06] TAVLI B., HEINZELMAN W., *Mobile Ad Hoc Networks: Energy-Efficient Real-Time Data Communications*, Springer, Netherlands, 2006.

[TIA 13] TIAN Y., MA J., PENG C., *et al.*, "A rational framework for secure communication", *Information Sciences*, vol. 250, pp. 215–226, 2013.

[TOB 80] TOBAGI F., HUNT V., "Performance analyses of carrier sense multiple access with collision detection", *Computer Networks*, vol. 4, no. 6, pp. 245–259, 1980.

[TOS 09] TOSIN A., "From generalized kinetic theory to discrete velocity modeling of vehicular traffic: a stochastic game approach", *Applied Mathematics Letters*, vol. 22, pp. 1122–1125, 2009.

[TOU 99] TOUTAIN L., *Réseaux Locaux et Internet*, Collection Réseaux et Télécommunications, Hermes Science, January, 1999.

[TUY 07] TUYLS K., PARSONS S., "What evolutionary game theory tells us about multiagent learning", *Artificial Intelligence*, vol. 171, pp. 406–416, 2007.

[UYA 14] UYANIK G.S. ABDEL RAHMAN M.J., KRUNZ M., "Optimal channel assignment with aggregation in multi-channel systems: a resilient approach to adjacent-channel interference", *Ad Hoc Networks*, vol. 20, pp. 64–76, 2014.

[VON 28] VON NEUMANN J., "Zur Theorie der Gesellschaftsspiele", *Mathematische Annalen*, vol. 100, 1928.

[VON 44] VON NEUMANN J., MORGENSTERN O., *Theory of Games and Economical Behaviour*, Princeton University Press, 1944.

[WAN 05] WANG X., KAR K., "Throughput modeling and fairness issues in CSMA/CA based ad-hoc networks", *INFOCOM, 24th annual joint conference of the IEEE Computer and Communications Societies*, vol. 1, pp. 23–34, 2005.

[WAN 09] WANG B., HAN Z., LIU K.J.R., "Distributed relay selection and power control for multiuser cooperative communication networks using Stackelberg game", *IEEE Transactions on Mobile Computing*, vol. 8, no. 7, p. 975, July 2009.

[WAN 10] WANG B., WU Y., RAY LIU K.J., "Game theory for cognitive radio networks: an overview", *Computer Networks*, vol. 54, pp. 2537–2561, 2010.

[WAR 52] WARDROP J., "Some theoretical aspects of road traffic research", *Proceedings of the Institution of Civil Engineers, Part II*, pp. 352–362, 1952.

[WAT 13] WATSON J., *Strategy, an Introduction to Game Theory W.W.*, Norton & Company, New York, 2013.

[WEI 95] WEIBULL, *Evolutionary Game Theory*, The MIT Press, 1995.

[WOO 01] WOO K, YU C., LEE D., *et al.*, "Non-blocking, localized routing algorithm for balanced energy consumption in mobile *ad hoc* networks", *International Symposium on Modeling, Analysis and Simulation of Computer and Telecommunication Systems (MASCOTS)*, pp. 117–124, 2001.

[WU 14] WU T.Y., LEE W.T., GUIZANI N., *et al.*, "Incentive mechanism for P2P files sharing based on social network and game theory", *Journal of Network and Computer Applications*, vol. 41, pp. 47–55, 2014.

[XIA 07] XIA W., QI Z., "Power control for cognitive radio base on game theory", *Wireless Communications , Networking and mobile Computing 2007, WiCom 2007, International conference on Networking and mobile Computing*, pp. 1256–1259, 2007.

[XIA 08] XIAO J., XIONG N., YANG L.T., *et al.*, "A joint selfish routing and channel assignment game in wireless mesh networks", *Computer Communications*, vol. 31, pp. 1447–1459, 2008.

[XIA 14] XIAOHUI N., FENG S., XUHAI H., *et al.*, "Predicting the protein solubility by integrating chaos games representation and entropy in information theory", *Expert Systems with Applications*, vol. 41, pp. 1672–1679, 2014.

[XU 12] XU S., HOU M., NIU K., *et al.*, "Coverage and capacity optimization in LTE network based on non-cooperative games", *The Journal of China Universities of Posts and Telecommunications*, vol. 19, no. 4, pp. 14–21, August 2012.

[YAN 10] YANG. J., SHI H., "Two Rounds Game CSMA Algorithm of WSNs", *International Conference on Networking and Digital Society (ICNDS)*, vol. 1, pp. 270-273, 2010.

[YAN 13] YANG L., KIM H., ZHANG J., *et al.*, "Pricing-based decentralized spectrum access control in cognitive radio networks" *IEEE/ACM Transactions on Networking*, vol. 21, no. 2, p. 522, April 2013.

[YE 02] YE W., HEIDEMANN J., ESTRIN D., "An energy-efficient MAC protocol for wireless sensor networks", *IEEE INFOCOM* 2002, *twenty first annual joint conference of the IEEE Computer and communications societies*, vol. 3, pp. 1567–1576, 2002.

[YU 04] YU Y., CAI X., GIANNAKIS G., "On the instability of slotted ALOHA with capture", *IEEE transactions on Wireless Communications*, vol. 5, no. 2, pp. 257–261, 2006.

[ZHA 00] ZHAO Q., TONG L., "Semi-blind collision resolution in random access wireless *ad hoc* networks", *IEEE Trans Signal Processing*, vol. 48, no. 10, pp. 2910–2920, 2000.

[ZHA 07] ZHAO L., ZHANG J., YANG K., *et al.*, "Using incompletely cooperative game theory in mobile *ad hoc* networks", *IEEE International conference on communications ICC'07*, pp. 3401–3406, 2007

[ZHA 09] ZHANG G., CONG L., ZHAO L., *et al.*, "Competitive resource sharing based on game theory in cooperative relay networks", *ETRI Journal*, vol. 31, no. 1, pp. 89–91, February 2009.

[ZHA 11] ZHANG Y., GUIZANI M., *Game Theory for Wireless Communications and Networking*, CRC Press Taylor Francis group, 2011.

[ZHA 12] ZHAI L.J., JI H., LI X., *et al.*, "Optimal resource allocation scheme for cognitive radio networks with relay selection based on game theory", *The Journal of China Universities of Posts and Telecommunications*, vol. 19, no. 6, pp. 25–28, December 2012.

[ZHA 13a] ZHANG Y., HUANG D., MIN J., *et al.*, "The evolution game analysis of clustering for asymmetrical multi-factors in WSNs", *Computers and Electrical Engineering*, vol. 39, pp. 1746–1757, 2013.

[ZHA 13b] ZHAO P., YANG X., WANG J., *et al.*, "Admission control on multipath routing in 802.11-based wireless mesh networks", *Ad Hoc Networks*, vol. 11, pp. 2235–2251, 2013.

[ZHE 14] ZHENG W., SU T., ZHANG H., *et al.*, "Distributed power optimization for spectrum-sharing femtocell networks: a fictitious game approach", *Journal of Network and Computer Applications*, vol. 37, pp. 315–322, 2014.

[ZHU 14] ZHU J., CHEN Y., SHEN Y., *et al.*, "Secrecy transmission capacity in noisy wireless *ad hoc* networks", *Ad Hoc Networks*, vol. 21, pp. 123–133, 2014.

[ZOR 97] ZORZI M., RAO R., "Energy constrained error control for wireless channels", *IEEE Personal Communications*, vol. 4, no. 6, pp. 27–33, 1997.

[ZOU 54] ZOU C., CHIGAN C., "On game theoretic DSA-driven MAC for cognitive radio networks", *Computer Communications*, vol. 32, pp. 1944–1954, 2009.

Index

A, B, C

associativity-based routing (ABR), 29
automatic repeat request (ARQ), 29, 33, 34
asynchronous transfer mode (ATM), 7
bit error ratio (BER), 55
busy tone, 31
clear channel detection (CCA), 18
code division multiple access (CDMA), 55
communication network, 35
Cournot duopoly, 53–54
cyclic redundancy check (CRC), 18, 23

D

distributed coordination function (DCF), 10, 16–19, 21
distributed inter frame space (DIFS), 18, 19, 25
destination sequenced distance vector (DSDV), 8, 29
direct sequence spread spectrum (DSSS), 16
duration, 23, 24
dynamic source routing (DSR), 8, 29

E, F

enhanced DCF (EDCF), 10, 93
erasure coding (ECIP), 74, 79, 102
extended IFS (EIFS), 18, 19
first input first output (FIFO), 61
forward error correction (FEC), 29, 33
freeway, 5
frequency division multiple access(FDMA), 55
frequency hopping spread spectrum (FHSS), 16

G, H, I

GAF protocol, 32, 33
game equilibrium, 57, 64, 84, 103

geographic adaptive fidelity (GAF), 32, 33
global state routing (GSR), 8
global system for mobile communication (GSM), 26
high frequency (HF), 26
IEEE 802.11, 16, 17, 28, 31
IFS, 18, 19, 20
infrared (IF), 16
inter frame space,

L, M

logical link control (LLC), 16, 17, 33, 34
long frame, 20–22, 95, 99
low earth orbit (LEO), 7
Lyapunov's stability theory, 58
Manhattan, 5
Markov chain, 64, 68, 69, 90, 96
MATLAB, 105
military, 1, 6
mixed strategy, 39, 50, 52
mobile *ad hoc* network (MANET), 5–10, 30, 33, 34, 37, 63, 65, 93, 110

N, O, P

Nash–Cournot equilibrium, 54
nodes
  exposed, 95, 99
  hidden, 20, 21, 95, 99, 100
optimized link state routing (OLSR), 8
orthogonal frequency division multiplexing (OFDM), 16
personal computer memory card international association (PCMCIA), 26
physical layer (PHY), 16, 17, 18, 19, 23, 25, 31, 56
Poisson process, 13, 65, 66, 68, 83, 101
power aware multi access protocol with signaling (PAMAS), 28, 31, 32
power-saving mode (PSM), 21, 28
prisoner's dilemma, 52
pure strategy, 39

Q, R

quality of service (QoS), 9, 17, 34, 55, 58, 62, 63, 74, 94
random walk, 5
received signal strength indicator (RSSI), 18
Reed–Solomon codes, 75
relative neighborhood graph (RNG), 3
RS, 75, 78, 86, 87

S, T, U

short IFS (SIFS), 18–20, 23–25
signal-stability adaptive routing (SSA), 29
SPAN protocol, 32, 33
switched port analyzer (SPAN), 32, 33
time division multiple access (TDMA), 55
uniqueness, 47, 49, 50
unit disk graph (UDG), 3
utility, 40

W, Z

Wardrop equilibrium, 57
wireless fidelity (WiFi), 17
wireless local area network (WLAN) 37,
wireless mesh network (WMN), 5–7
wireless routing protocol (WRP), 8
wireless sensor network (WSN), 5–7
zone routing protocol (ZRP), 9

Other titles from

iSTE

in

Networks and Telecommunications

2014

BATTU Daniel
New Telecom Networks: Enterprises and Security

BEN MAHMOUD Mohamed Slim, GUERBER Christophe, LARRIEU Nicolas, PIROVANO Alain, RADZIK José
Aeronautical Air–Ground Data Link Communications

BITAM Salim, MELLOUK Abdelhamid
Bio-inspired Routing Protocols for Vehicular Ad-Hoc Networks

CAMPISTA Miguel Elias Mitre, RUBINSTEIN Marcelo Gonçalves
Advanced Routing Protocols for Wireless Networks

CHETTO Maryline
Real-time Systems Scheduling 1: Fundamentals
Real-time Systems Scheduling 2: Focuses

EXPOSITO Ernesto, DIOP Codé
Smart SOA Platforms in Cloud Computing Architectures

MELLOUK Abdelhamid, CUADRA-SANCHEZ Antonio
Quality of Experience Engineering for Customer Added Value Services

OTEAFY Sharief M.A., HASSANEIN Hossam S.
Dynamic Wireless Sensor Networks

PEREZ André
Network Security

PERRET Etienne
Radio Frequency Identification and Sensors: From RFID to Chipless RFID

REMY Jean-Gabriel, LETAMENDIA Charlotte
LTE Standards
LTE Services

TANWIR Savera, PERROS Harry
VBR Video Traffic Models

VAN METER Rodney
Quantum Networking

XIONG Kaiqi
Resource Optimization and Security for Cloud Services

2013

ASSING Dominique, CALÉ Stéphane
Mobile Access Safety: Beyond BYOD

BEN MAHMOUD Mohamed Slim, LARRIEU Nicolas, PIROVANO Alain
Risk Propagation Assessment for Network Security: Application to Airport Communication Network Design

BEYLOT André-Luc, LABIOD Houda
Vehicular Networks: Models and Algorithms

BRITO Gabriel M., VELLOSO Pedro Braconnot, MORAES Igor M.
Information-Centric Networks: A New Paradigm for the Internet

BERTIN Emmanuel, CRESPI Noël
Architecture and Governance for Communication Services

DEUFF Dominique, COSQUER Mathilde
User-Centered Agile Method

DUARTE Otto Carlos, PUJOLLE Guy
Virtual Networks: Pluralistic Approach for the Next Generation of Internet

FOWLER Scott A., MELLOUK Abdelhamid, YAMADA Naomi
LTE-Advanced DRX Mechanism for Power Saving

JOBERT Sébastien *et al.*
Synchronous Ethernet and IEEE 1588 in Telecoms: Next Generation Synchronization Networks

MELLOUK Abdelhamid, HOCEINI Said, TRAN Hai Anh
Quality-of-Experience for Multimedia: Application to Content Delivery Network Architecture

NAIT-SIDI-MOH Ahmed, BAKHOUYA Mohamed, GABER Jaafar, WACK Maxime
Geopositioning and Mobility

PEREZ André
Voice over LTE: EPS and IMS Networks

2012

AL AGHA Khaldoun
Network Coding

BOUCHET Olivier
Wireless Optical Communications

DECREUSEFOND Laurent, MOYAL Pascal
Stochastic Modeling and Analysis of Telecoms Networks

DUFOUR Jean-Yves
Intelligent Video Surveillance Systems

EXPOSITO Ernesto
Advanced Transport Protocols: Designing the Next Generation

JUMIRA Oswald, ZEADALLY Sherali
Energy Efficiency in Wireless Networks

KRIEF Francine
Green Networking

PEREZ André
Mobile Networks Architecture

2011

BONALD Thomas, FEUILLET Mathieu
Network Performance Analysis

CARBOU Romain, DIAZ Michel, EXPOSITO Ernesto, ROMAN Rodrigo
Digital Home Networking

CHABANNE Hervé, URIEN Pascal, SUSINI Jean-Ferdinand
RFID and the Internet of Things

GARDUNO David, DIAZ Michel
Communicating Systems with UML 2: Modeling and Analysis of Network Protocols

LAHEURTE Jean-Marc
Compact Antennas for Wireless Communications and Terminals: Theory and Design

RÉMY Jean-Gabriel, LETAMENDIA Charlotte
Home Area Networks and IPTV

PALICOT Jacques
Radio Engineering: From Software Radio to Cognitive Radio

PEREZ André
IP, Ethernet and MPLS Networks: Resource and Fault Management

TOUTAIN Laurent, MINABURO Ana
Local Networks and the Internet: From Protocols to Interconnection

2010

CHAOUCHI Hakima
The Internet of Things

FRIKHA Mounir
Ad Hoc Networks: Routing, QoS and Optimization

KRIEF Francine
Communicating Embedded Systems / Network Applications

2009

CHAOUCHI Hakima, MAKNAVICIUS Maryline
Wireless and Mobile Network Security

VIVIER Emmanuelle
Radio Resources Management in WiMAX

2008

CHADUC Jean-Marc, POGOREL Gérard
The Radio Spectrum

GAÏTI Dominique
Autonomic Networks

LABIOD Houda
Wireless Ad Hoc and Sensor Networks

LECOY Pierre
Fiber-optic Communications

MELLOUK Abdelhamid
End-to-End Quality of Service Engineering in Next Generation Heterogeneous Networks

PAGANI Pascal *et al.*
Ultra-wideband Radio Propagation Channel

2007

BENSLIMANE Abderrahim
Multimedia Multicast on the Internet

PUJOLLE Guy
Management, Control and Evolution of IP Networks

SANCHEZ Javier, THIOUNE Mamadou
UMTS

VIVIER Guillaume
Reconfigurable Mobile Radio Systems

Printed and bound by CPI Group (UK) Ltd, Croydon, CR0 4YY

13/09/2026

14961016-0008